D0986859

ADVANCES IN CELL AND MOLECULAR BIOLOGY

Volume 3

CONTRIBUTORS

G. F. Bahr
Joseph Bryan
John H. Frenster
Robert T. Johnson
P. M. Larsen
Ben Lung
Marilyn A. Masek
Shirley L. Nakatsu
Potu N. Rao
Leslie Wilson

ADVANCES IN CELL AND MOLECULAR BIOLOGY

Edited by E. J. DuPraw

STANFORD UNIVERSITY
SCHOOL OF MEDICINE
STANFORD, CALIFORNIA

VOLUME 3—1974

ACADEMIC PRESS New York San Francisco London
A Subsidiary of Harcourt Brace Jovanovich, Publishers

ACADEMIC PRESS, INC.
111 Fifth Avenue, New York, New York 10003

United Kingdom Edition published by
ACADEMIC PRESS, INC. (LONDON) LTD.
24/28 Oval Road, London NW1

LIBRARY OF CONGRESS CATALOG CARD NUMBER: 73-163767

ISBN 0-12-008003-6

PRINTED IN THE UNITED STATES OF AMERICA

CONTENTS

Induction of Chromosome Condensation in Interphase Cells

Potu N. Rao and Robert T. Johnson

Structural "Bands" in Human Chromosomes

G. F. Bahr and P. M. Larsen

LIST OF CONTRIBUTORS

Numbers in parentheses indicate the pages on which the authors' contributions begin.

G. F. BAHR (191), Armed Forces Institute of Pathology, Washington, D.C.

JOSEPH BRYAN (21), Department of Biology, University of Pennsylvania, Philadelphia, Pennsylvania

JOHN H. FRENSTER (1), Division of Medical Oncology, Department of Medicine, Stanford University School of Medicine, Stanford, California

ROBERT T. JOHNSON (135), Department of Zoology, University of Cambridge, Cambridge, England

P. M. LARSEN (191), Armed Forces Institute of Pathology, Washington, D.C.

BEN LUNG (73), Department of Anatomy, Stanford University School of Medicine, Stanford, California

MARILYN A. MASEK (1), Division of Medical Oncology, Department of Medicine, Stanford University School of Medicine, Stanford, California

SHIRLEY L. NAKATSU (1), Division of Medical Oncology, Department of Medicine, Stanford University School of Medicine, Stanford, California

POTU N. RAO (135), Department of Developmental Therapeutics, The University of Texas M. D. Anderson Hospital and Tumor Institute, at Houston, Houston, Texas

LESLIE WILSON (21), Department of Pharmacology, Stanford University School of Medicine, Stanford, California

ADVANCES IN CELL AND MOLECULAR BIOLOGY

Volume 3

ULTRASTRUCTURAL PROBES OF DNA TEMPLATES WITHIN HUMAN BONE MARROW AND LYMPH NODE CELLS*

John H. Frenster,
Shirley L. Nakatsu,
and Marilyn A. Masek

DIVISION OF MEDICAL ONCOLOGY
DEPARTMENT OF MEDICINE
STANFORD UNIVERSITY SCHOOL OF MEDICINE
STANFORD, CALIFORNIA

I. Introduction

In differentiated cells, only a fraction of the DNA within each cell is active as a template for RNA synthesis (Frenster *et al.*, 1963), and

* Supported in part by Grants CA-10174, AM-01006, and CA-13524 from the National Institutes of Health and by a Research Scholar Award from the Leukemia Society to Dr. Frenster.

this fraction may be characteristic for each different type of tissue (Paul and Gilmour, 1968).

Control of DNA template activity is evident in many biological systems. During the course of spermatogenesis, a progressive fall in DNA template activity and RNA synthesis results in the formation of mature sperm synthesizing almost no RNA (Ringertz *et al.*, 1970). When such a mature sperm hybridizes with a mature ovum during fertilization, the sperm DNA templates are derepressed, allowing an equal maternal and paternal contribution to gene expression in all resulting cells of the embryo and the adult (Davidson, 1968). Similarly, during the course of embryogenesis in the developing liver, a progressive decrease in the diversity of RNA species being synthesized is observed, but when the adult liver is induced to regenerate following partial hepatectomy, derepression of previously repressed DNA templates again allows the reappearance of those RNA species characteristic of the embryonic state (Church and McCarthy, 1967).

Increasing derepression of previously repressed DNA templates also is noted if neoplastic cells are assayed for the diversity of RNA species being synthesized as the neoplasm progresses first to a benign nodule, then to a spontaneous neoplasm, and finally to a transplantable highly malignant neoplasm (Turkington, 1971). Such a derepression of normally repressed DNA templates also characterizes human leukemic lymphocytes (Neiman and Henry, 1969; Sawada *et al.*, 1973), and may account for the reappearance within adult human neoplasms of cell surface antigens characteristic of normal fetal life (Gold, 1971). A quite similar reappearance of normal fetal antigens has been observed in both chemically induced and virus-induced neoplasms in experimental animals (Herstein and Frenster, 1972).

These findings suggest that variable DNA template activity and its control are central to molecular and cellular events occurring during embryogenesis, organ regeneration, and neoplasia and have prompted the development of high resolution techniques for assessing DNA template activity within single intact cells (Frenster, 1971).

II. Acridine Orange Binding to DNA Templates

A. Within Isolated DNA

Acridine orange is a planar polycyclic molecule which binds with high affinity to isolated DNA molecules (Rigler, 1969). The binding of acridine orange to DNA involves at least two physical binding modes: (a) a stacking interaction which involves the intercalation of acridine orange molecules between adjacent base pairs within the interior of the DNA helix and which occurs at low ratios of ligand to nuleic acid (Lerman, 1963); and (b) an electrostatic interaction between the basic groups of the acridine orange molecule and the acidic phosphate groups on the exterior of the DNA helix, which becomes prominent at high ratios of ligand to nucleic acid (Mason and McCaffery, 1964).

B. Within Isolated Chromatin

The prior presence of chromatin proteins, such as polycationic histones, on the DNA helix effectively decreases the reactivity of such DNA to acridine orange (Rigler, 1969). The mechanisms of such restriction of binding of acridine orange are: (a) the prevention by polycationic histones of local strand separations within the DNA helix, which are necessary to allow intercalation of acridine orange into the helix (Frenster, 1965b); and (b) the neutralization by polycationic histones of phosphate groups on the exterior of the DNA helix, which otherwise would be available for reaction with the basic groups of the acridine orange molecule (Frenster, 1969). As a consequence of such inhibition of acridine orange binding to DNA by histones, acridine orange microfluorescent probes (Rigler, 1969) have been used to distinguish chromatin states in which histones are tightly bound to underlying DNA helices from those in which histones are loosely bound to DNA (Frenster, 1965a); this technique distinguishes DNA templates that are inactive or active, respectively, in RNA synthesis (Frenster *et al.*, 1963; Frenster, 1965a).

C. Within Isolated Cells

Fluorescent acridine orange probes have made it possible to detect changes in DNA template activity in a wide variety of cell systems by means of microspectrofluorimetry. These cell systems include the increases in DNA template activity that occur in lymphocytes after phytohemagglutinin stimulation (Killander and Rigler, 1969), in nucleated erythrocytes after cell hybridization (Bolund *et al.*, 1969), and in lymphocytes obtained from patients with infectious mononucleosis (Bolund *et al.*, 1970). In addition, the same method has been used to detect decreases in DNA template activity occurring in differentiating spermatozoa during spermatogenesis (Ringertz *et al.*, 1970) and in cells cultured at high cell densities (Zetterberg and Auer, 1970).

When used in such microspectrofluorimetric analyses of single fixed cells, acridine orange probes can distinguish single-stranded nucleic acid binding sites from double-stranded sites by physical means (Rigler, 1969), but cannot distinguish DNA binding sites from RNA binding sites by chemical means. Because of the increasing evidence for the natural occurrence of both double-stranded RNA–RNA duplexes (Jelinek and Darnell, 1972) and of single-stranded DNA loops (Saucier and Wang, 1972), this low chemical specificity becomes a limiting factor in such analyses.

Another limiting factor, the low resolution of separate binding sites possible with fluorescent light microscopy, suggested the need for development of a high-resolution electron microscopic technique for detecting acridine orange binding sites specific for DNA. It was also desirable that such techniques might be employed on section tissue so that *in vivo* topological relations between cells could be preserved for the DNA template analysis.

III. Ultrastructural Probes of DNA Templates

A. Within Bone Marrow Cells

A technique has been developed for the aspiration of living bone marrow cells from healthy subjects or untreated patients (Frenster, 1971). The cells are obtained in bone marrow spicules which preserve the *in vivo* topological relationships between the cells. When such living bone marrow spicules are allowed to react with acridine orange and DNase (Frenster, 1972) and are then examined by high-resolution electron microscopic techniques, an electron-dense reaction product is localized over the sites of active DNA templates within the euchromatin portion of the cell nucleus (Frenster, 1971). The intensity of individual probe site accumulation can be analyzed further by measuring the size of the individual reaction products observed within single cells. Such individual probe site accumulations range between 0.025 and 0.1 μ in diameter (B probes), 0.1–0.35 μ in diameter (A probes), and > 0.35 μ in diameter (AA probes) and represent the reaction product formed between acridine orange and osmic acid (Nakatsu *et al.*, 1974).

When the number of probe sites was counted within each of 123 individual differentiating granulocytes (Fig. 1) from normal living

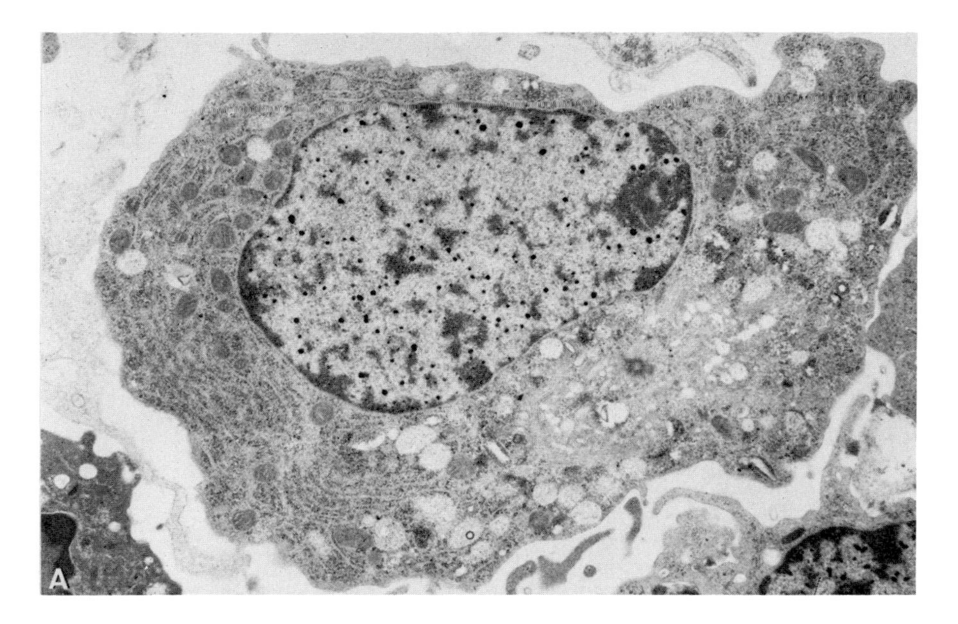

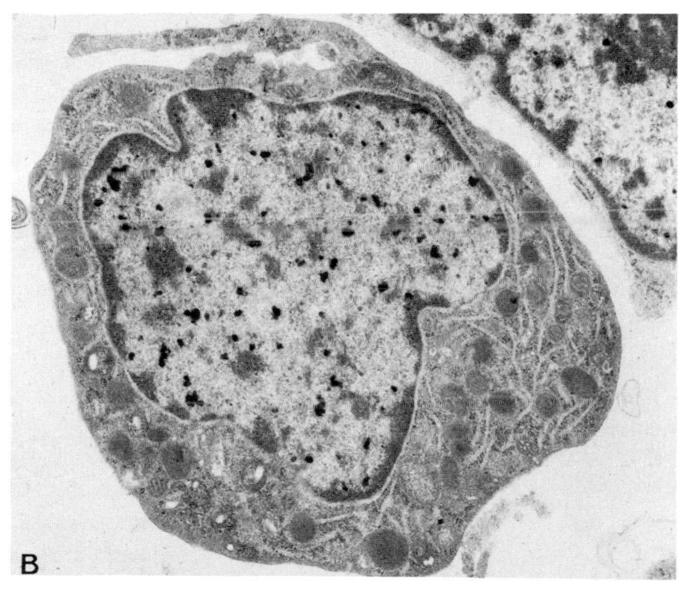

FIG. 1. Probe analysis of DNA templates within differentiating granulocytes in normal human bone marrow. Electron-dense reaction products are localized exclusively within the euchromatin portion of the cell nucleus. (A) Promyelocyte. A nucleolus is present in the nucleus. Three types of immature granules are seen in the cytoplasm, together with a centriole, a well-developed Golgi apparatus, and a significant amount of endoplasmic reticulum. ×6500. (B) Myelocyte. Only two types of granules are seen in the cytoplasm. ×6500.

TABLE I
PROBE SITE COUNTS WITHIN NORMAL DIFFERENTIATING GRANULOCYTES
(N = 123)

Cells	Percent cells containing		Mean site count/ positive cell	
	A Probes	B Probes	A Probes	B Probes
Promyelocytes	91.5	100.	9.55	35.5
Myelocytes	85.9	96.4	8.75	35.1
Metamyelocytes	47.6	85.7	3.9	36.3
Band granulocytes	0	20.0	0	8.0
Segmented granulocytes	0	12.2	0	10.1

human bone marrow, it was found that the incidence of cells positive for either A or B probes declined as cell differentiation progressed (Table I). In addition, the number of A probes per positive cell declined as cell differentiation progressed (Table I). These probe site data correlate well with the decline in RNA synthesis previously noted within such differentiating cells (Feinendegan *et al.*, 1964), and they indicate that a progressive restriction of the number and activity of DNA templates occurs as a feature of normal granulocytic differentiation (Nakatsu *et al.*, 1974).

Similarly, when the number of probe sites was determined within each of 189 individual differentiating erythrocytes (Fig. 2) from normal living human bone marrow, it was found that the incidence of cells positive for either A or B probes declined as cell differentiation progressed (Table II). In addition, the number of A probes per positive cell declined as cell differentiation progressed (Table II). These probe site data correlate well with the decline in RNA synthesis previously noted within such differentiating cells (Feinendegan *et al.*, 1964), and indicate that a progressive restriction of the number and activity of DNA templates occurs as a feature of normal erythrocytic differentiation.

When the number of probe sites within each of 97 individual mononuclear cells (Figs. 3–6) from normal living human bone marrow was determined (Table III), it was found that cells active in phagocytosis (monocytes, macrophages, endothelial cells, reticulum cells) had a consistently higher incidence of cells positive for either A or B probes than did nonphagocytic cells (lymphocytes, plasma cells). In addition, the number of A probes per positive cell was higher for phagocytic

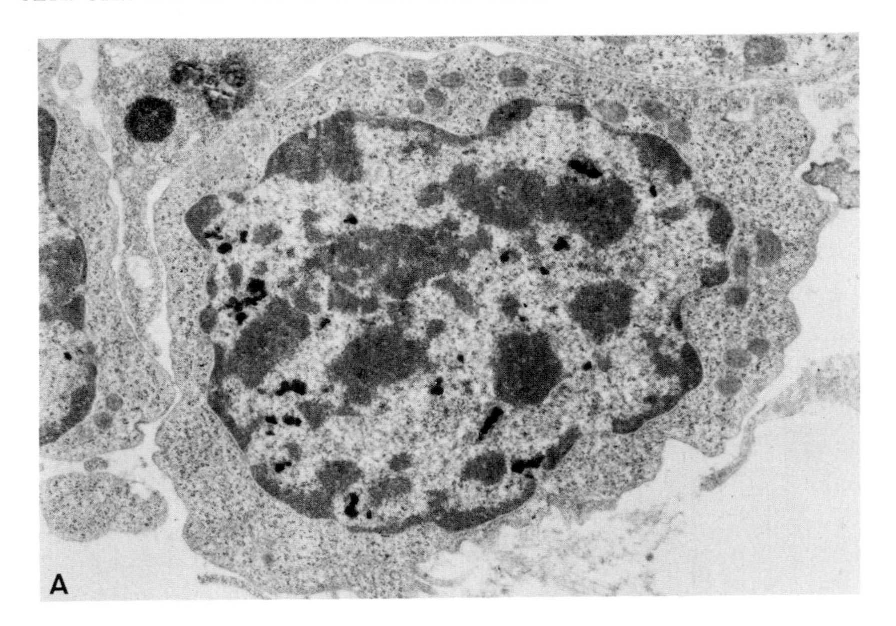

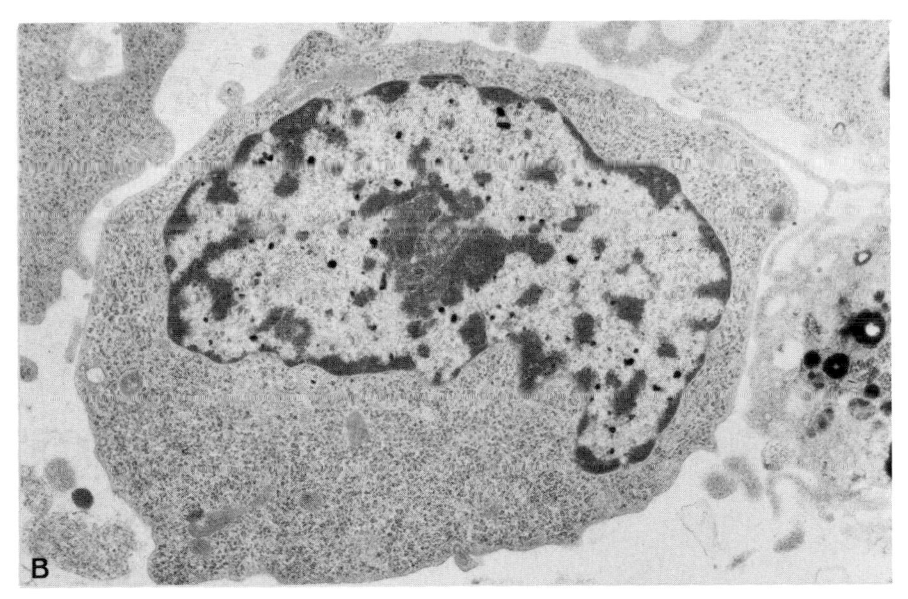

FIG. 2. Probe analysis of DNA templates within differentiating erythrocytes in normal human bone marrow. Electron-dense reaction products are localized exclusively within the euchromatin portion of the cell nucleus. (A) Proerythroblast. The cytoplasm contains numerous polyribosomes and fine ferritin granules, with evidence of rhopheocytosis at the plasma membrane. ×6500. (B) Erythroblast. The cytoplasm contains numerous polyribosomes, but lesser degrees of ferritin or rhopheocytosis. ×6500.

TABLE II
PROBE SITE COUNTS WITHIN NORMAL DIFFERENTIATING ERYTHROCYTES
(N = 189)

Cells	Percent cells containing		Mean site count/positive cell	
	A Probes	B Probes	A Probes	B Probes
Proerythroblasts	100	100	16.0	8.4
Early erythroblasts	84.7	91.6	8.5	16.3
Late erythroblasts	14.8	17.2	3.4	11.8
Nucleated erythrocytes	0	8.3	0	12.0

cells than for nonphagocytic cells (Table III). These probe site data suggest that phagocytic cells possess more numerous and more active DNA templates than do nonphagocytic cells in normal human bone marrow. By contrast, small numbers of undifferentiated cells (? stem cells; Fig. 4B) had very high incidences of either A or B probes, and a high number of A probes per positive cell, suggesting that such undifferentiated marrow cells have very numerous and very active DNA templates in normal human bone marrow.

B. WITHIN LYMPH NODE LYMPHOCYTES

A technique has been developed for the surgical removal and analysis of living lymph node cells from untreated patients with Hodgkin's disease (Archibald and Frenster, 1973). The cells are obtained in cubes of lymph node aliquots which preserve the *in vivo* topological relationships between the cells. When such lymph node aliquots are fixed in glutaraldehyde (Archibald and Frenster, 1973), then reacted with acridine orange and DNase, and finally examined by high resolution electron microscopic techniques, an electron-dense reaction product is localized over the sites of active DNA templates within the euchromatin portion of the cell nucleus (Frenster, 1972).

Lymph node lymphocytes in Hodgkin's disease are believed to represent host immunoactive cells directed against neoplastic Reed-Sternberg cells (Archibald and Frenster, 1973). When the number of probe sites was determined within each of 194 individual immunoactive lymphocytes (Fig. 7) from intact lymph nodes of untreated patients with Hodgkin's

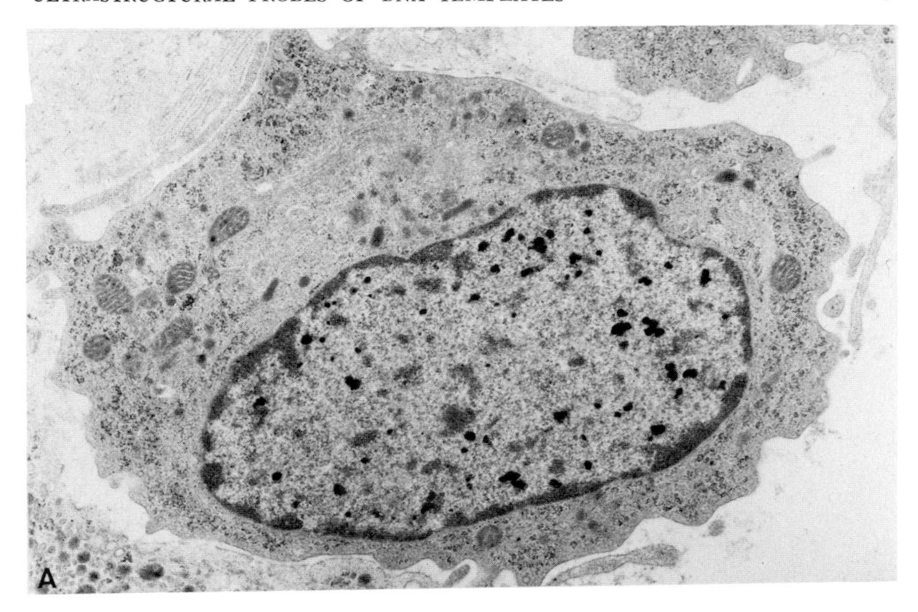

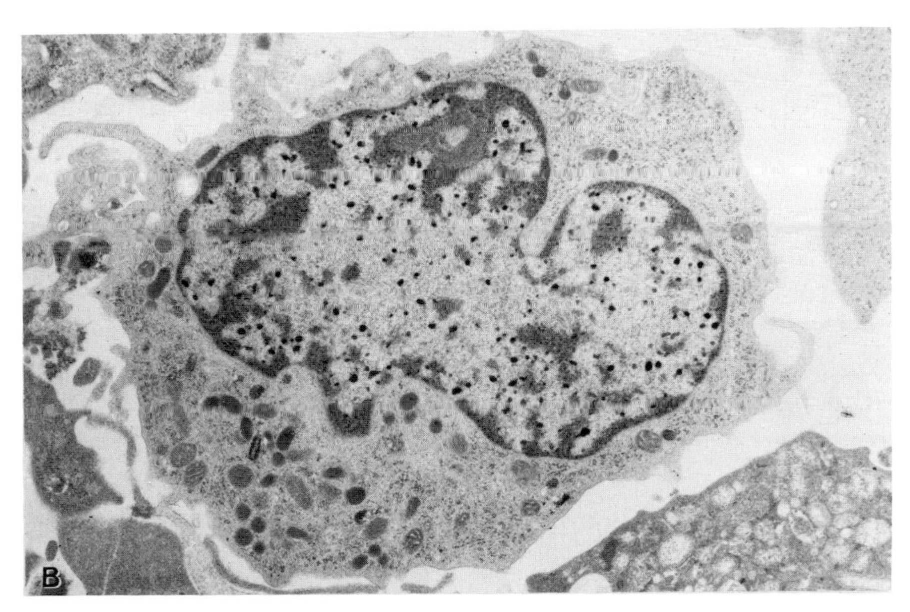

FIG. 3. Probe analysis of DNA templates within differentiating monocytes in normal human bone marrow. Electron-dense reaction products are localized exclusively within the euchromatin portion of the cell nucleus. (A) Early monocyte. The cytoplasm contains a centriole, a well-developed Golgi apparatus, and a small number of dense granules. ×6500. (B) Late monocyte. The cytoplasm contains short segments of rough endoplasmic reticulum and an increasing number of dense granules. ×6500.

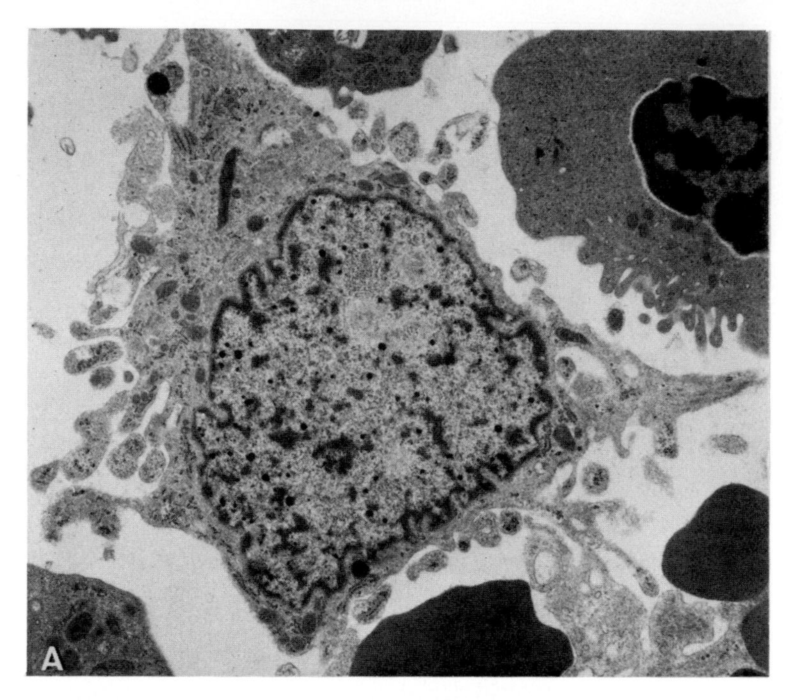

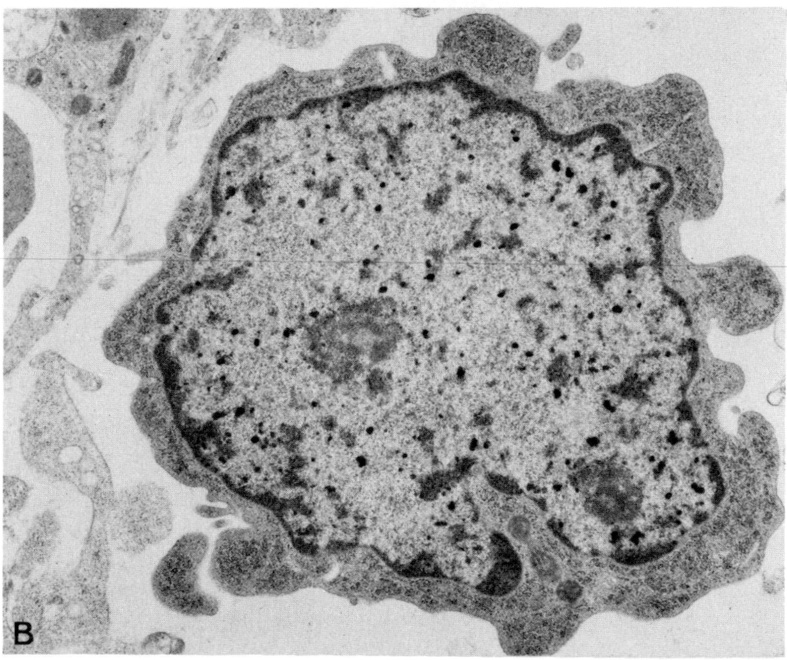

Fig. 4. Probe analysis of DNA templates within mononuclear cells in normal human bone marrow. Electron-dense reaction products are localized exclusively within the euchromatin portion of the cell nucleus. (A) Macrophage. The plasma

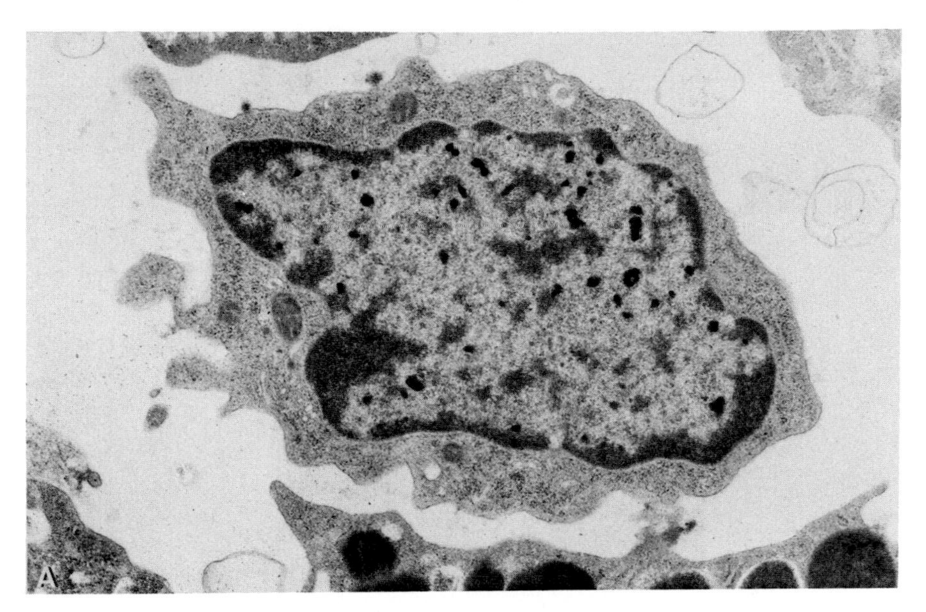

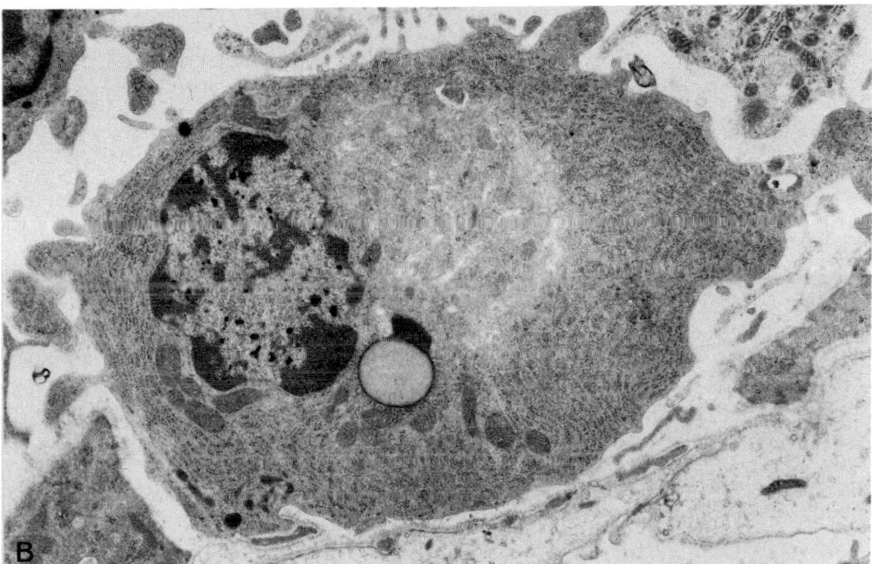

FIG. 5. Probe analysis of DNA templates within lymphoid cells of normal human bone marrow. Electron-dense reaction products are localized exclusively within the euchromatin portion of the cell nucleus. (A) Small lymphocyte. The cytoplasm contains monoribosomes but no dense granules or endoplasmic reticulum. ×13,000. (B) Plasma cell. The cytoplasm contains a large amount of rough endoplasmic reticulum, a homogeneous inclusion body, and a large Golgi area. ×3250.

---

membrane is highly convoluted, and the cytoplasm contains numerous short segments of rough endoplasmic reticulum. ×3750. (B) Undifferentiated cell. The cytoplasm is extremely scanty and contains only an occasional mitochondrion. ×7500.

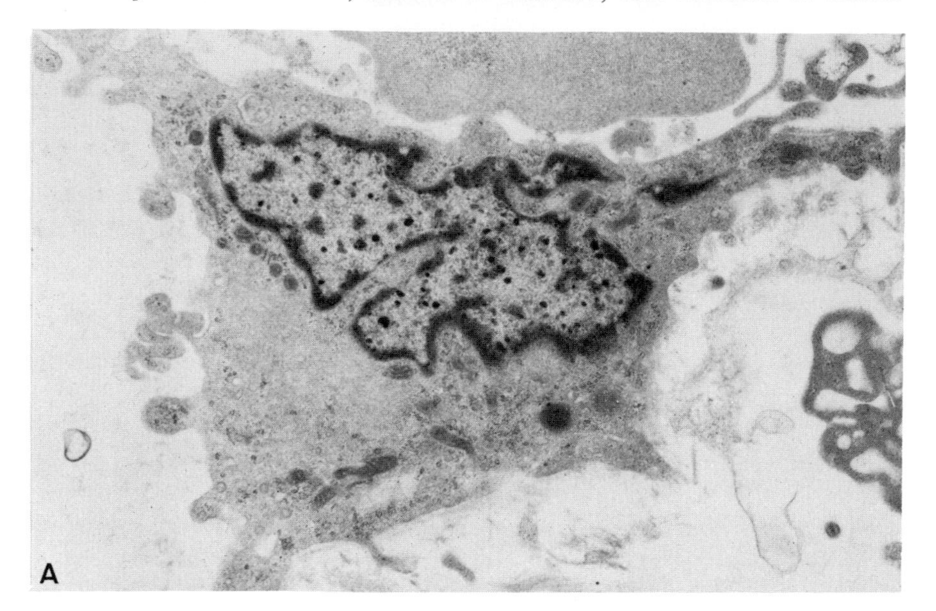

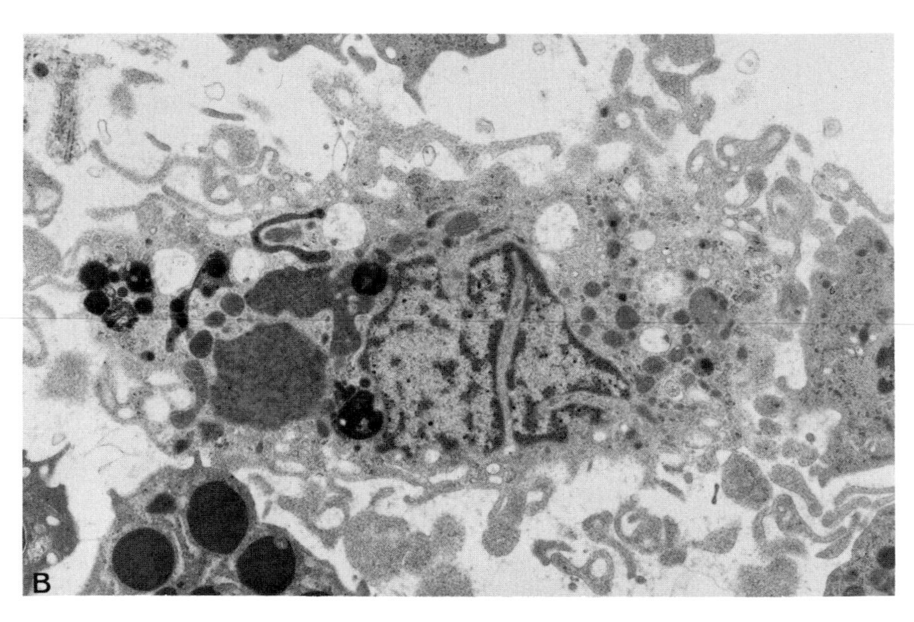

FIG. 6. Probe analysis of DNA templates within reticuloendothelial cells of normal human bone marrow. Electron-dense reaction products are localized exclusively within the euchromatin portion of the cell nucleus. (A) Endothelial cells. The endothelial cell is lining a sinus (above) and displays an active plasma membrane toward the surrounding marrow cord (below). ×3250. (B) Reticulum cell. The cytoplasm contains numerous phagosomes and ferritin particles, and the plasma membrane is extremely convoluted. ×3250.

TABLE III
PROBE SITE COUNTS WITHIN NORMAL MARROW MONONUCLEAR CELLS ($N = 97$)

Cells	Percent cells containing		Mean site count/ positive cell	
	A Probes	B Probes	A Probes	B Probes
Monocytes	77.3	100.0	7.3	43.4
Macrophages	100.0	100.0	11.1	17.4
Endothelial cells	90.0	100.0	8.8	16.8
Reticulum cells	52.5	81.2	12.2	64.3
Plasma cells	46.2	84.6	5.5	26.5
Lymphocytes	11.1	44.4	4.0	20.5

disease, it was found that the incidence of cells positive for either A or AA probes increased with the incidence of cells activated from monoribosomal lymphocytes to polyribosomal lymphocytes (Table IV). In addition, the number of AA probes per positive cell increased as lymphocyte activation increased (Table IV). These probe site data correlate well with the increase in polyribosomes (Tokuyasu *et al.*, 1968) and in acridine orange binding (Killander and Rigler, 1969) previously noted within such lymphocytes during activation, and indicate that a progressive increase in the number and activity of DNA templates occurs as a feature of lymphocyte immune activation (Stanley *et al.*, 1971; Frenster and Rogoway, 1970).

C. WITHIN DIVIDING CELLS

When probe sites were analyzed within a number of large nonlymphocytic polyribosomal cells in the lymph nodes of untreated patients with Hodgkin's disease (Fig. 8), it was found that the number of A or AA probe sites per positive cell decreases as these cells enter the various mitotic phases of cell division. The reaction products within mitotic cells (Fig. 8B) are confined to noncondensed areas of the reconstituting nucleus and correlate with the previous finding that only a small amount of RNA is synthesized by dividing cells (Jakob, 1972). This suggests that a significant restriction in the number and activity of DNA templates occurs as a feature of metaphase and anaphase during mitotic division (Keshgegian *et al.*, 1971; Nakatsu *et al.*, 1974).

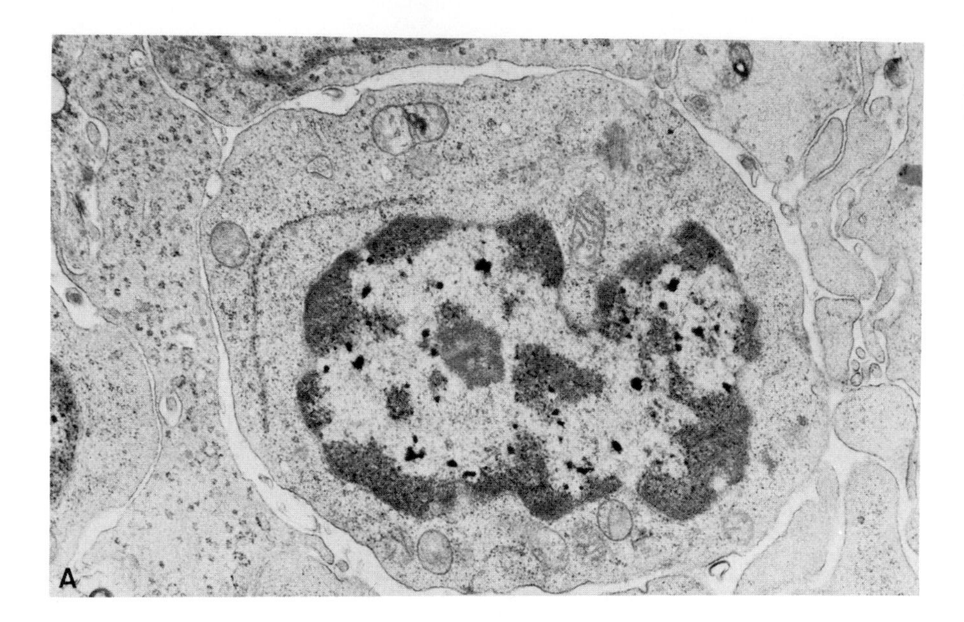

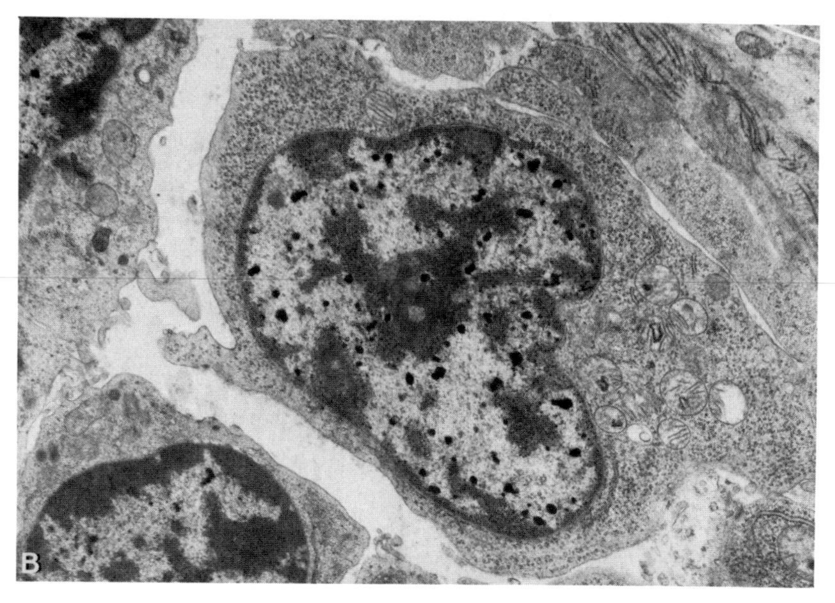

FIG. 7. Probe analysis of DNA templates within lymphocytes of a positive lymph node taken from an untreated patient with Hodgkin's disease, nodular sclerosis type. Electron-dense reaction products are localized exclusively within the euchromatin portion of the cell nucleus. (A) Monoribosomal lymphocyte. The cytoplasm contains a centriole, and the ribosomes are largely arrayed in a monosomal manner. ×13,000. (B) Polyribosomal lymphocyte. The cytoplasmic ribosomes are largely arrayed in polysomal clusters. ×9750.

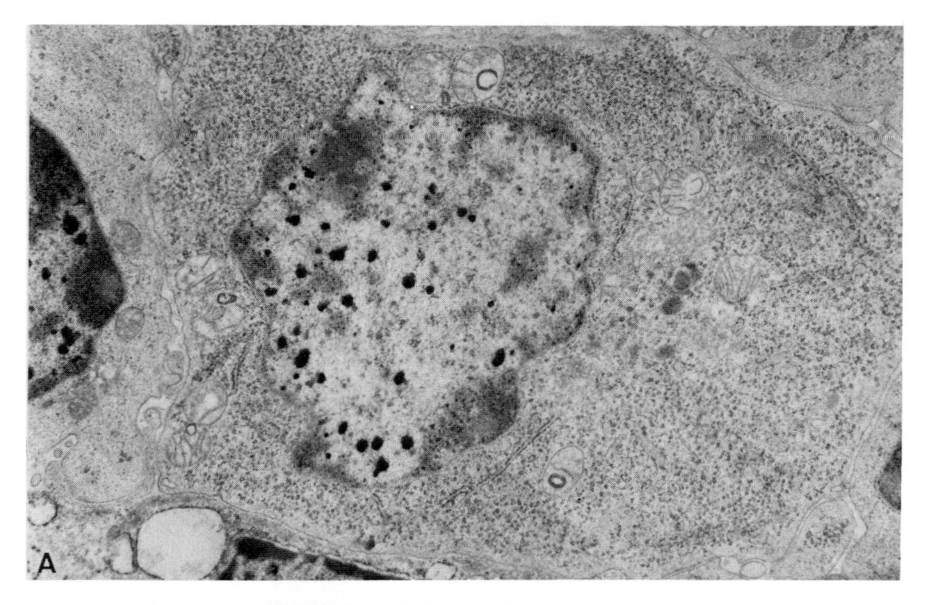

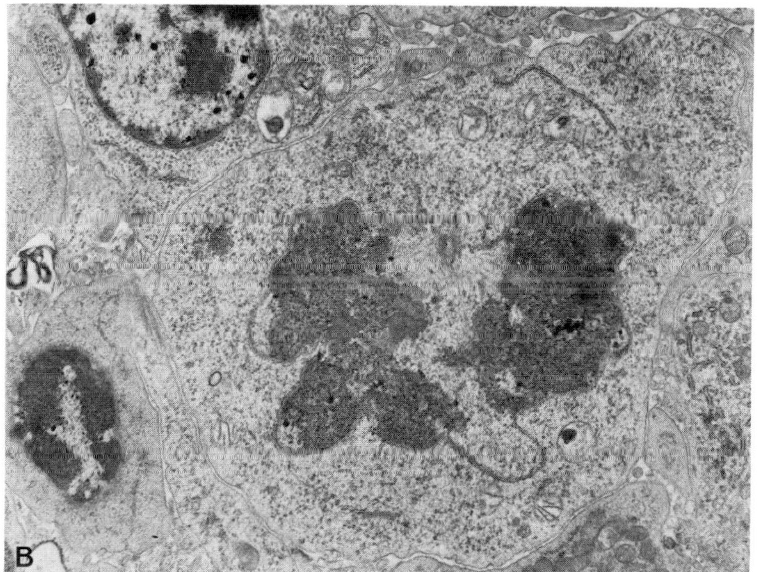

FIG. 8. Probe analysis of DNA templates within polysomal cells of a positive lymph node taken from an untreated patient with Hodgkin's disease, nodular sclerosis type. (A) Large polysomal cell with electron-dense reaction products localized exclusively within the euchromatin portion of the cell nucleus. Aside from clusters of polyribosomes, the cytoplasm contains a rare strand of rough endoplasmic reticulum and small numbers of dense granules near the Golgi apparatus. ×6500. (B) Large polysomal cell in early telophase of mitosis, with condensed chromosomes, centriole with mitotic spindle, and early reconstitution of the nuclear membrane. The electron-dense reaction products are localized to the euchromatin portion of the cell nucleus between the masses of condensed chromosomes. The cytoplasm contains clusters of polyribosomes and a rare strand of rough endoplasmic reticulum. ×6500.

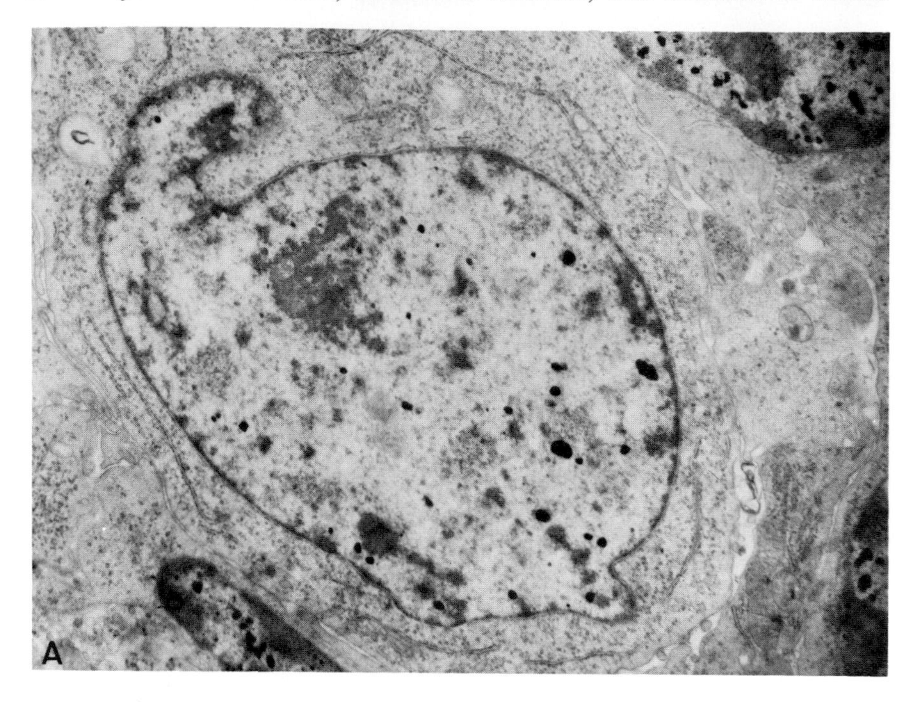
A

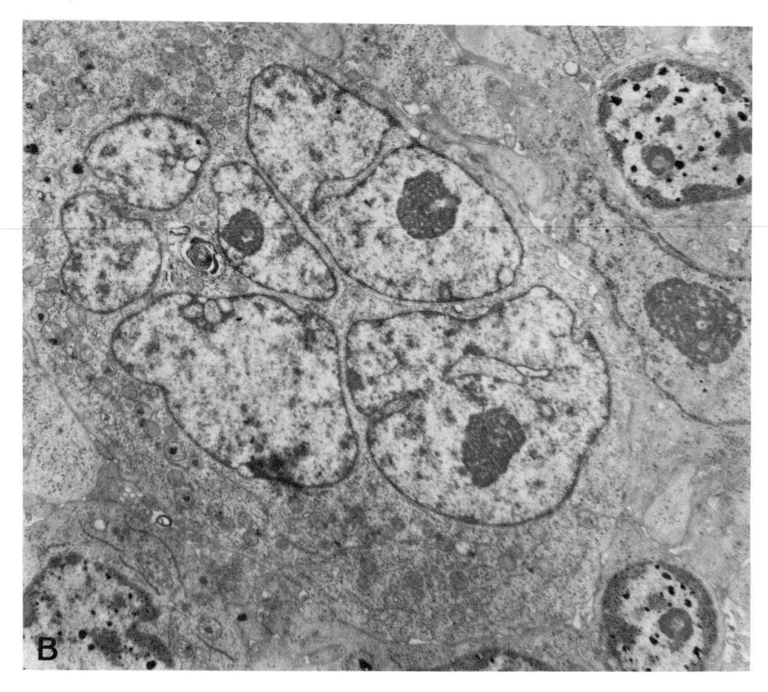
B

TABLE IV
PROBE SITE COUNTS WITHIN HODGKIN'S DISEASE LYMPHOCYTES
($N = 194$)

	Percent cells containing		Mean site count/ positive cell	
Lymphocytes	AA Probes	A Probes	AA Probes	A Probes
Monoribosomal	53.2	94.8	2.7	18.1
Mixed	67.8	100.0	3.2	18.7
Polyribosomal	79.5	100.0	3.5	17.1

D. Within Neoplastic Cells

When the probe sites within a number of mononuclear or polynuclear Reed-Sternberg cells (Bernhard and Leplus, 1964; Mori and Lennart, 1969) in the lymph nodes of untreated patients with Hodgkin's disease were examined (Fig. 9), it was found that the incidence of mononuclear Reed-Sternberg cells positive for either A or AA probes was quite high (Fig. 9A), while a significant number of polynuclear Reed-Sternberg cells were found to be completely negative for A or AA probes. These probe site data correlate well with the decreased incidence of DNA synthesis observed in polynuclear Reed-Sternberg cells as compared to mononuclear Reed-Sternberg cells (Peckham and Cooper, 1969), and suggest that during the course of neoplastic cell survival a significant fraction of advanced neoplastic cells leave the proliferative cycle by virtue of decreasing their DNA template activity (Saunders and Mauer, 1969; Peckham and Cooper, 1969; Ahearn and Trujillo, 1972).

FIG. 9. Probe analysis of DNA templates within neoplastic cells of a positive lymph node taken from an untreated patient with Hodgkin's disease, nodular sclerosis type. (A) Large mononuclear Reed-Sternberg cell, with electron-dense reaction products localized exclusively within the euchromatin portion of the cell nucleus. The nucleus is irregularly shaped, invaginated, and contains a large nucleolus, while the cytoplasm is largely polyribosomal with an occasional strand of rough endoplasmic reticulum. ×7500 (B) Polynuclear Reed-Sternberg cell, showing almost no reaction products, while surrounding lymphocytes are heavily labeled by reaction products. Each of the lobes of the nucleus are themselves invaginated and at least three large nucleoli are evident. The cytoplasm is polyribosomal with an occasional thin strand of rough endoplasmic reticulum. ×3750.

IV. Other Ligands to DNA

A number of direct ligands to DNA consist of small molecules that are potentially capable of being utilized as probes of DNA primary, secondary, or tertiary structure (Frenster, 1965b). They can be divided into ligands binding preferentially to single-stranded or to double-stranded DNA, and by means of such preference effecting either increases or decreases of RNA synthesis when added to sensitive cells (Frenster, 1965b). None so far has achieved the utility, high-resolution, and specificity of acridine orange, but studies in their possible use are continuing (Frenster and Herstein, 1973).

V. Summary

Gene expression via RNA synthesis requires RNA precursor substrates, RNA polymerases, and active DNA templates for effective transcription of genetic information. In many biological systems, restriction of DNA template activity, as detected by fluorescent or ultrastructural probes of active DNA template sites, presents a controlling factor in gene expression and RNA synthesis.

REFERENCES

Ahearn, M. J., and Trujillo, J. M. (1972). *Proc. Amer. Ass. Cancer Res.* **13**: 108.

Archibald, R. B., and Frenster, J. H. (1973). *Monogr. Nat. Cancer Inst.* **36**; 239.

Bernhard, W., and Leplus, R. (1964). "Fine Structure of the Normal and Malignant Lymph Node." Macmillan, New York.

Bolund, L., Ringertz, N. R., and Harris, H. (1969). *J. Cell Sci.* **4**: 71.

Bolund, L., Gahrton, G., Killander, D., Rigler, R., and Wahren, B. (1970). *Blood* **35**: 322.

Church, R. B., and McCarthy, B. J. (1967). *J. Mol. Biol.* **23**: 477.

Davidson, E. H. (1968). "Gene Activity in Early Development." Academic Press, New York.

Feinendegan, L. E., Bond, V. P., Cronkite, E. P., and Hughes, W. L. (1964). *Ann. N.Y. Acad. Sci.* **113**: 737.

Frenster, J. H. (1965a). *Nature (London)* **206**: 680.

Frenster, J.H. (1965b). *Nature (London)* **208**: 1093.

Frenster, J. H. (1969). *In* "Handbook of Molecular Cytology" (A. Lima-de-Faria, ed.), pp. 251–276. North-Holland Publ., Amsterdam.

Frenster, J. H. (1971). *Cancer Res.* **31**: 1128.

Frenster, J. H. (1972). *Nature (London), New Biol.* **236**: 175.

Frenster, J. H., and Herstein, P. R. (1973). *N. Engl. J. Med.* **228**: 1224.

Frenster, J. H., and Rogoway, W. M. (1970). *In* "Proceedings of the Fifth Leukocyte Culture Conference" (J. E. Harris, ed.), pp. 359–371. Academic Press, New York.
Frenster, J. H., Allfrey, V. G., and Mirsky, A. E. (1963). *Proc. Nat. Acad. Sci. U.S.* **50**: 1026
Gold, P. (1971). *Annu. Rev. Med.* **22**: 85.
Herstein, P. R., and Frenster, J. H. (1972). *In* "Embryonic and Fetal Antigens in Cancer" (N. G. Anderson and J. H. Coggins, eds.), Vol. 2, pp. 5–7. Nat. Tech. Inform. Serv., U.S. Dep. Commerce, Springfield, Virginia.
Jakob, K. M. (1972). *Exp. Cell Res.* **72**: 370.
Jelinek, W., and Darnell, J. E. (1972). *Proc. Nat. Acad. Sci. U.S.* **69**: 2537.
Keshgegian, A. A., Meisner, L. F., and Frenster, J. H. (1971). *In* Proceedings of the Fourth Leukocyte Culture Conference" (O. R. McIntyre, ed.), pp. 361–366. Appleton, New York.
Killander, D., and Rigler, R. (1969). *Exp. Cell Res.* **54**: 163.
Lerman, L. S. (1963). *Proc. Nat. Acad. Sci. U.S.* **49**: 94.
Mason, S. F., and McCaffery, A. J. (1964). *Nature (London)* **204**: 468.
Mori, Y., and Lennart, K. (1969). "Electron Microscopic Atlas of Lymph Node Cytology and Pathology." Springer-Verlag, Berlin and New York.
Nakatsu, S. L., Masek, M. A., Landrum, S., and Frenster, J. H. (1974). *Nature (London)* **248**: 334.
Neiman, P. E., and Henry, P. H. (1969). *Biochemistry* **8**: 275.
Paul, J., and Gilmour, R. S. (1968). *J. Mol. Biol.* **34**: 305.
Peckham, M. J., and Cooper, E. H. (1969). *Cancer* **24**: 135.
Rigler, R. (1969). *Ann. N.Y. Acad. Sci.* **157**: 211.
Ringertz, N. R., Gledhill, B. L., and Darzynkiewicz, Z. (1970). *Exp. Cell Res.* **62**: 204.
Saucier, J. M., and Wang, J. C. (1972). *Nature (London), New Biol.* **239**: 167.
Saunders, E. F., and Mauer, A. M. (1969). *J. Clin. Invest.* **48**: 1299.
Sawada, H., Gilmore, V. H., and Saunders, G. F. (1973). *Cancer Res.* **33**: 428.
Stanley, D. A., Frenster, J. H., and Rigas, D. A. (1971). *In* "Proceedings of the Fourth Leukocyte Culture Conference" (O. R. McIntyre, ed.), pp. 1–11. Appleton, New York.
Tokuyasu, K., Madden, S. C., and Zeldis, L. J. (1968). *J. Cell Biol.* **39**: 630.
Turkington, R. W. (1971). *Cancer Res.* **31**: 427.
Zetterberg, A., and Auer, G. (1970). *Exp. Cell Res.* **62**: 262.

BIOCHEMICAL AND PHARMACOLOGICAL PROPERTIES OF MICROTUBULES

Leslie Wilson

DEPARTMENT OF PHARMACOLOGY
STANFORD UNIVERSITY SCHOOL OF MEDICINE
STANFORD, CALIFORNIA

Joseph Bryan

DEPARTMENT OF BIOLOGY
UNIVERSITY OF PENNSYLVANIA
PHILADELPHIA, PENNSYLVANIA

I. Introduction

A. Aim of This Review

The antimitotic drugs colchicine, colcemid, vinblastine sulfate, and vincristine sulfate have become valuable chemical tools for investigating the functional roles and biochemical properties of microtubules. For a complete account of the occurrence, ultrastructure, and functions of microtubules, the reader is referred to the excellent reviews of Porter (1966), Tilney (1971), and Olmsted and Borisy (1973). Our aim in this article is to discuss in detail the use of the antimitotic drugs for elucidating the mechanisms of drug action, together with the biochemical and functional properties of microtubules.

B. Occurrence and Function of Microtubules

Microtubules are long tubelike structures approximately 220–250 Å in diameter, which appear to exist in all eukaryotic cells. They can be classified arbitrarily into two general categories, *stable* and *labile*. The stable microtubules are found primarily in cilia and flagella, where they are organized into the familiar $9 + 2$ array. In contrast, microtubules found in the axons and dendrites of the central nervous system, in the mitotic apparatus of dividing cells, and in the cytoplasm both of animal and plant cells fall into the labile category. Before the introduction of glutaraldehyde fixation for electron microscopy by Sabatini *et al.* (1963), only stable microtubules could be observed with the electron microscope because harsher fixation treatments (e.g., osmium tetroxide) destroyed the labile microtubules. However, after this significant advance, efforts of many investigators have demonstrated the ubiquitous nature of microtubules and their occurrence in unusually large numbers in some tissues (e.g., in the central nervous system).

It is generally believed that labile microtubules are in a state of "dynamic equilibrium" (see Inoué and Sato, 1967). The essential feature of this model is that molecules of the protein *tubulin* (the microtubule building block) occur in a cytoplasmic pool which is in equilibrium with the assembled microtubules. The model implies that a cell is able to control the polymerization and depolymerization of these units. Various characteristics of labile microtubules fit this model rather well: for example, they are quickly destroyed (or depolymerized) at low temperatures, under high hydrostatic pressure, or by the action of antimitotic drugs. In contrast, stable microtubules do not appear to be in equilibrium with a tubulin pool; they can be isolated intact from cilia and flagella, they do not depolymerize at low temperature or under high pressure, and they are not destroyed by any of the chemical agents that destroy labile microtubules (Stephens, 1971; Behnke and Forer, 1967; Tilney and Gibbons, 1968).

Microtubules evidently participate in a number of different cellular processes, and it is likely that the differences in stability among microtubules of various origin are related to their different cellular roles. The many similarities, as well as the differences in the biochemical and drug binding of various microtubule proteins, will be discussed in later sections.

II. Biochemical Properties of Microtubule Proteins (*Tubulins*)

A. Subunit Structure

Early investigations on the chemical properties of microtubule subunits were carried out with proteins obtained both from stable outer doublet microtubules of *Tetrahymena* cilia (Renaud *et al.*, 1968) and from sea urchin sperm tails (summarized in Stephens, 1971). Intact microtubules were isolated by a series of differential extractions that removed the flagellar plasma membrane and various matrix proteins. For *Tetrahymena* cilia, acetone powders of the purified microtubules were then extracted with low ionic strength solutions following procedures similar to those used in the isolation of muscle actin.

By these means, a protein was solubilized which had a sedimentation coefficient of 6.0 S and an apparent molecular weight of 104,000 (determined by sedimentation equilibrium). Addition of denaturing agents (8 *M* urea or 6 *M* guanidine·HCl) reduced the apparent molecular weight to 55,000, which suggested that the 104,000 molecular weight protein was composed of two subunits of similar size. Polyacrylamide gel electrophoresis of this material in 8 *M* urea (after reduction and acetylation) revealed the presence of a single, or two closely spaced, protein bands.

The subunit structure of outer-doublet microtubules from the sea urchin *Strongylocentrotus droebachiensis* was investigated in a similar fashion (Stephens, 1968). Treatment of purified outer doublet microtubules with Salyrgan, an organic mercurial, solubilized a 5.1 S protein with an apparent molecular weight of 130,000 (by sedimentation equilibrium). Addition of denaturing agents resulted in dissociation of the protein to 3 S subunits having an apparent molecular weight of 59,000. Dissociation occurred either in the presence or in the absence of mercaptoethanol, suggesting that the subunits were not held together by disulfide bonds. Similar studies were carried out with subunits isolated from outer doublet microtubules of S. *purpuratus* sperm tails (Shelanski and Taylor, 1968).

In these early reports, heavy emphasis was placed upon the marked similarities between muscle actin and the proteins isolated from microtubules. Both appeared to possess similar amino acid composition and molecular sizes and contained bound nucleotides. However, after the apparent homology between actin and tubulin was investigated in detail by Stephens (1970a), it became clear that tubulin and actin differ in molecular size and amino acid sequence (suggested by peptide mapping) and are distinct proteins.

A second approach to the investigation of microtubule subunit composition involved the use of radioactive colchicine to isolate a high-affinity colchicine-binding receptor protein (see also Section IV,B). Such a receptor had been found earlier in several eukaryotic tissues (Borisy and Taylor, 1967a,b; Wilson and Friedkin, 1967), and was purified from porcine brain extracts by Weisenberg *et al.* (1968). This colchicine receptor complex, and similar colchicine complexes that were purified subsequently from several other sources, all had sedimentation coefficients of 6 S, with an apparent molecular weight of 120,000 (by sedimentation equilibrium or gel filtration; Wilson and Friedkin, 1967; Weisenberg *et al.*, 1968; Bryan and Wilson, 1971). Denaturing agents dissociated the complexes to 55,000–60,000 molecular weight subunits. For a time, the colchicine binding protein was thought to be a dimer composed of two identical 60,000 molecular weight subunits. Considerable circumstantial evidence (reviewed in Weisenberg *et al.*, 1968) suggested that this 120,000 molecular weight colchicine receptor, which was found in the soluble fraction of a large number of eukaryotic cells and tissues, corresponded to tubulin.

1. *Possible Heterogeneity of Microtubules*

The possibility of microtubule heterogeneity was suggested by Behnke and Forer (1967), who found that microtubules from several sources had different solubility properties when exposed to different temperatures or when treated with proteolytic enzymes. Later, Stephens (1970b) demonstrated that the proteins from the A and B subfibers* of sea urchin

* Each microtubule in an outer doublet microtubule pair is commonly referred to as a subfiber.

outer doublet microtubules could be solubilized differentially by use of a thermal fractionation technique. The initial earlier evidence suggested that each subfiber might be composed of a single protein. On the basis of amino acid composition, solubility, and mobility on polyacrylamide gels, Stephens proposed that subfiber A was composed of one protein (tubulin A) and subfiber B of a different protein (tubulin B).

In *Chlamydomonas,* the molecular composition of flagellar outer-doublet microtubules was similarly explored by Jacobs and McVittie (1970), who used more refined electrophoretic techniques. They also concluded that one subfiber was composed of a single protein subunit (α), while a second protein subunit (β) was derived from the B subfiber or the central pair microtubules (for further discussion, see Section IV).

2. *Heterogeneity of Tubulin*

However, a different type of heterogeneity has recently been proposed by Bryan and Wilson (1971). Highly purified colchicine-receptor (tubulin) from chick embryo brain was found to be composed of two different protein subunits, which could be separated on 8 *M* urea polyacrylamide gels after reduction and acetylation (Fig. 1). The slower migrating component was termed α; the faster component, β.

SDS polyacrylamide gel electrophoresis of the two subunits under *high ionic strength, neutral pH conditions* in the gel (Shapiro *et al.*, 1967; Weber and Osborn, 1969) showed only a single component, which had an apparent molecular weight of 55,000. Since all attempts to subfractionate the tubulin into α:α tubulin and β:β tubulin were unsuccess-

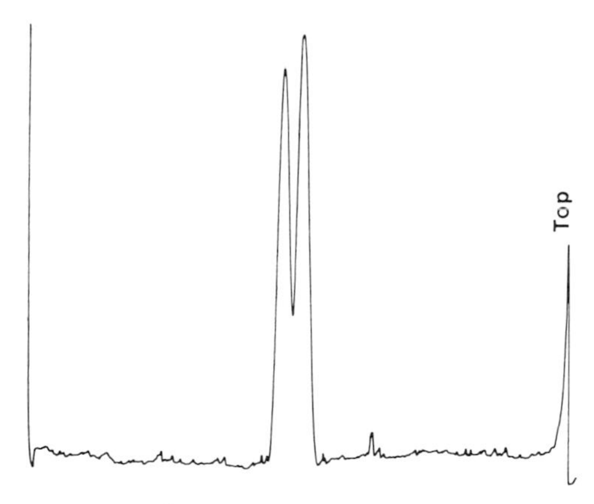

FIG. 1. Separation of α and β subunits of chick embryo brain tubulin by urea polyacrylamide gel electrophoresis. (Data of Bryan and Wilson, 1971. Reproduced with permission of National Academy of Sciences.)

ful, it was suggested that tubulin may be a heterodimer (α:β); this idea implies that cytoplasmic microtubules (composed of the colchicine-binding heterodimers) are heteropolymers (Bryan and Wilson, 1971). The same model was proposed independently by Feit *et al.* (1971) and later by Fine (1971). In addition, Feit *et al.* (1971) reinvestigated the subunit composition of sea urchin sperm tail outer doublet microtubules (see Section II,A,1), and demonstrated that α and β subunits are present in both the A and B subfibers.

Evidence for tubulin heterogeneity was subsequently extended to a number of other systems, including sea urchin egg mitotic apparatus tubulin (Bibring and Baxandall, 1971), sea urchin egg vinblastine-induced crystals (Bryan, 1972b), sea urchin outer doublet microtubules (Meza *et al.*, 1972; Meza, 1972), *Tetrahymena* cilia (Everhart, 1971), and *Chlamydomonas* flagellar outer doublet microtubules (Witman *et al.*, 1972a,b). Although there are some differences among these systems, the existence of subunit heterogeneity is regarded as well established. The chemical differences between the α and β subunits have not been completely defined. Differences have been reported in their amino acid compositions (Table I) (Bryan and Wilson, 1971; Fine, 1971; Meza, 1972; Eipper, 1972; Witman *et al.*, 1972a,b), in peptide maps and cyanogen bromide fragments (Fine, 1971; Meza, 1972; Feit *et al.*, 1971), and of their N-terminal amino acid sequences (Table II; Luduena and Woodward, 1972).

Early evidence obtained by sodium dodecyl sulfate (SDS) polyacrylamide gel electrophoresis suggested that the α and β subunits were of similar molecular size, since both subunits had the same mobility on the gels. More recently, however, the use of different SDS polyacrylamide gel electrophoresis systems has produced conflicting results. Two distinct protein bands have been obtained on some gels, suggesting that the subunits also differ in molecular size (Feit *et al.*, 1971; Fine, 1971; Witman *et al.*, 1972a,b; Meza *et al.*, 1972).

Cooke and Bryan (unpublished) have studied the mobilities of tubulin subunits on SDS polyacrylamide gels at different pH and ionic strength conditions. They found that the lower the ionic strength of the buffer in the gel, the greater is the *apparent* molecular weight difference between the two chains (Fig. 2A); the *higher* the pH, the *greater* the *apparent* molecular weight difference (Fig. 2B). Only the slower moving α subunit exhibits these pH and ionic strength-dependent mobility changes. The mobilities of the β subunit and of commonly used globular protein markers (e.g., bovine serum albumin, ovalbumin, β-lactoglobulin) are not affected by the pH or ionic strength changes. The β subunit has a constant apparent molecular weight of 52,000.

TABLE I
AMINO ACID COMPOSITIONS OF α AND β SUBUNITS OF TUBULINS FROM SEVERAL SOURCES[a]

Amino acid	Chick embryo brain[b]		Sea urchin egg, vinblastine crystals[c]		Sea urchin sperm tails, outer doublet microtubules[d]	
	α	β	α	β	α	β
Lys	16 (0.8)	16 (1.0)	16 (0.8)	17 (1.3)	17	19
His	11 (0.6)	9 (0.4)	8 (1.7)	8 (0.9)	9	9
Arg	20 (1.4)	20 (0.8)	20 (1.3)	18 (0.4)	16	18
CMCys	10 (0.8)	8 (0.4)	10 (0.9)	7 (0.9)	7	5
Asp	43 (1.5)	46 (3.5)	47 (2.9)	48 (1.7)	41	42
Thr	28 (0.7)	28 (0.7)	29 (1.7)	26 (1.7)	25	27
Ser	26 (0.4)	31 (0.8)	28 (2.1)	33 (1.3)	28	33
Glu	56 (1.7)	53 (9.0)	59 (2.5)	61 (2.6)	61	63
Pro	20 (2.0)	20 (2.5)	19 (3.3)	18 (0.9)	14	13
Gly	37 (1.4)	38 (1.1)	47 (1.3)	46 (1.7)	46	46
Ala	32 (1.1)	29 (1.0)	34 (1.7)	30 (1.3)	32	27
Val	33 (1.5)	31 (0.8)	31 (2.9)	30 (0.9)	31	27
Met	7 (0.8)	10 (1.2)	5 (1.7)	9 (2.6)	12	17
Ile	22 (0.6)	18 (0.6)	17 (3.3)	16 (1.7)	17	14
Leu	30 (1.3)	31 (0.9)	29 (1.3)	31 (0.4)	29	26
Tyr	15 (0.8)	15 (1.4)	14 (1.7)	13 (0.4)	14	12
Phe	19 (1.0)	20 (0.5)	17 (1.3)	18 (1.3)	15	15

[a] All proteins were reduced and carboxymethylated. Values denote residues per 55,000 MW; those in parentheses represent the variation at the 95% confidence level (data from 5 or 6 determinations on each chain).

[b] From Bryan and Wilson (1971).

[c] From Bryan (1972b).

[d] From Meza (1972).

Either of two mechanisms could account for the behavior of the α subunit: (1) differential SDS binding; or (2) residual electrostatic interactions within the SDS–protein complex. While these data can explain the disparity in published values for the molecular weights of the tubulin subunits, they leave us with a larger problem: What is the true molecular weight of the α subunit? If the molecular weights of the two subunits differ, then the amino acid compositions may be different from those originally calculated. The results also suggest that experiments designed to test directly the validity of the heterodimer hypothesis, by use of protein cross-linking reagents, will have to be interpreted with care, since the pH and ionic strength mobility dependence of the α subunit may be altered by cross-linking.

TABLE II
N-Terminal Amino Acid Sequences of α and β Subunits of Sea Urchin Outer Doublet Microtubules[a]

Residue No.	α Subunit	β Subunit
1	Met	Met
2	Arg	Arg
3	Glu	Glu
4	Ser (?)	Ile
5	Ile	Val
6	Ser (?)	His
7	Ile	Met
8	His	Glx
9	Val	Ala

[a] Data of Luduena and Woodward (1972).

B. Carbohydrate and Nucleotide Content of Tubulin

1. *Carbohydrates*

Falxa and Gill (1969) purified a 70,000 mw glycoprotein from calf brain, which they considered to be microtubule protein; this material contained approximately 1% carbohydrate (as glucose). Later, Goodman *et al.* (1971) reported that purified bovine brain tubulin contains approximately 1.2% carbohydrate. Margolis *et al.* (1972) analyzed the carbohydrate content of purified porcine brain tubulin and detected glucosamine, galactosamine, galactose, mannose, fucose, and sialic acid. Their work suggested the presence of both alkali-labile O-glycosidic linkages of *N*-acetylgalactosamine to serine and/or threonine, and alkali-stable linkages. Calculations indicated that at least two types of oligosaccharides were present and that their tubulin contained seven monosaccharide residues per mole.

In contrast with the previous results, Eipper (1972) found that highly purified tubulin actually contains little or no covalently linked carbohydrate. Eipper modified the tubulin purification procedure of Weisenberg *et al.* (1968) by substituting a pyrophosphate buffer for the phosphate buffer used originally. This buffer supposedly removed carbohydrate and nucleic acid contaminants which were associated with the tubulin prepared by the original method. Tubulin purified in this manner contains no covalently bound amino sugars and no more than 0.2% neutral sugars (1.2 moles of neutral sugar per mole of dimer, calibrated as

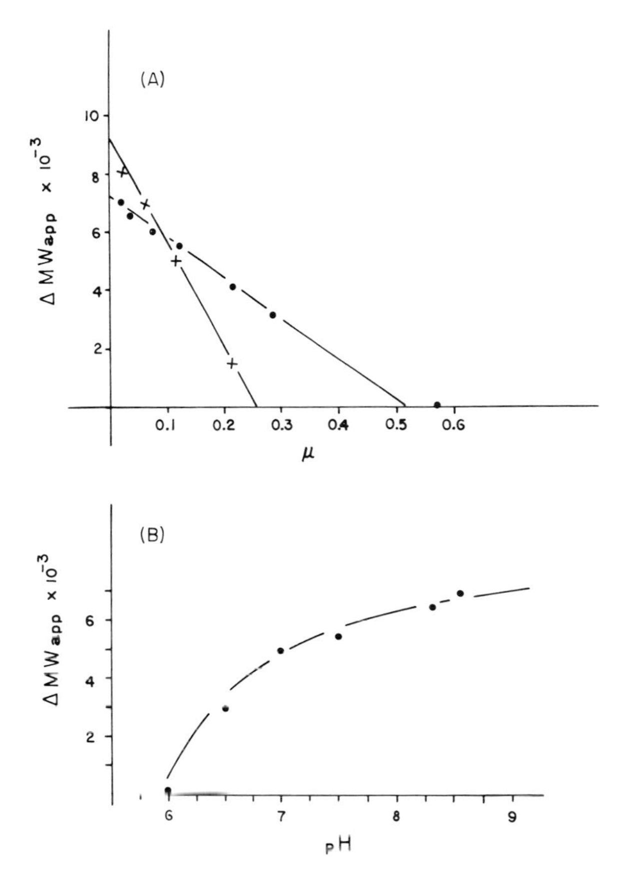

FIG. 2. Apparent molecular weight differences between the α and β subunits on SDS polyacrylamide gels under various conditions of ionic strength and pH. (A) Variation of ionic strength (●, Tris-glycinate buffer; ×, sodium phosphate buffer). (B) Variation of the pH.

galactose). Modification of the purification procedure did not affect the mobilities of the α and β subunits on polyacrylamide gels.

2. *Nucleotides*

Outer doublet microtubules from sea urchin flagella and from *Tetrahymena* cilia contain tightly associated guanine nucleotides (GTP, GDP, GMP) (Stephens *et al.*, 1967; Yanagisawa *et al.*, 1968; Shelanski and Taylor, 1968). The presence of guanine nucleotides has also been demonstrated in purified bovine brain tubulin (Weisenberg *et al.*, 1968). Brain tubulin is reported to contain 0.5 to 0.8 mole of a mixture of GDP and GTP per 120,000 gm of protein. The guanine nucleotides are tightly bound to the protein and do not exchange with exogenous GTP. However, a second GTP binding site has also been found, and GTP which is bound at this second site can exchange with exogenously added GTP.

Tubulin in sea urchin egg, vinblastine-induced crystals also possesses two guanine nucleotide binding sites (Bryan, 1972a). In crystals, both binding sites are occupied with *nonexchangeable* guanine nucleotide. Berry and Shelanski (1972) have suggested that there may be a transfer of the terminal phosphate of GTP in the exchangeable site of soluble brain tubulin to GDP in the nonexchangeable site. They have proposed that dephosphorylation occurs during polymerization of tubulin and have presented evidence that the assembled microtubule has a GTP:GDP ratio of 1. However, definitive experiments demonstrating dephosphorylation of GTP during polymerization have not yet been obtained. Recent developments permitting polymerization of microtubules *in vitro* (Section II,D) should make it possible to test the above hypothesis directly.

C. Phosphorylation of Tubulin

Goodman *et al.* (1970), influenced by reports that colchicine inhibits secretion in several systems in which cyclic adenosine monophosphate (cAMP) is thought to play a role, investigated the possible phosphorylation of tubulin by a cAMP-dependent protein kinase. They added bovine brain tubulin (purified according to Weisenberg *et al.*, 1968) to a protein kinase preparation purified from bovine brain and found that serine residues on the tubulin were phosphorylated. In addition, they also found that the tubulin preparations possessed intrinsic protein kinase activity, which could be stimulated by cAMP and could also phosphorylate serine residues on the tubulin. The enzymatic properties of the intrinsic neurotubule protein kinase were similar to those of a brain protein kinase studied by Miyamoto *et al.* (1969). However, neurotubule kinase was insensitive to Ca^{2+}, while the Miyamoto enzyme was inhibited by Ca^{2+}. Colchicine had no effect on the neurotubule kinase activity.

In a subsequent report, DiBella *et al.* (1971) indicated that the kinase activity could be separated from the tubulin by precipitation of the tubulin with magnesium ions. In contrast with these results, Soifer (1972a) reported that precipitation of tubulin by vinblastine did not separate the intrinsic protein kinase activity from the tubulin. The sensitivity to cAMP of the protein kinase activity associated with porcine brain tubulin (but not the activity itself) was destroyed by lyophilization (Soifer, 1972b). Also the degree of sensitivity of the kinase activity to cAMP also depended upon the protein substrate which was used. Although the specific activity of the vinblastine-precipitated kinase activity was the same as it was before vinblastine precipitation, it did exhibit greater sensitivity to cAMP. Kinase activity in vinblastine-precipitated tubulin was not influenced by addition of colchicine or vinblastine.

Lagnado *et al.* (1972) have demonstrated that bovine brain tubulin can be phosphorylated *in vitro,* and have extended this observation to tubulins from several other sources. Only the β subunit is phosphorylated. However, in these experiments the values of ^{3}H-cAMP binding to tubulin were low (1–4 mmoles of bound ^{3}H-labeled cAMP per mole of 58,000 MW monomer). Under conditions of maximal cAMP binding, ^{32}P incorporation into protein phosphoserine seldom exceeded 20 mmoles per mole of monomer. Eipper (1972) has also shown that phosphorylation of rat brain tubulin can occur *in vivo.* The tubulin, which was labeled with $^{32}PO_4$ *in vivo* and then purified, contained 0.8 ± 0.2 mole of phosphate per mole of *dimer.* All the labeled phosphate was covalently linked to the β subunit.

These results demonstrate convincingly that phosphorylation of tubulin can occur in cells, but the mechanism by which this is accomplished and its significance remain unclear. The fact that only low levels of cAMP binding to tubulin preparations have been detected, and the indication that the kinase activity may be separable from tubulin (DiBella *et al.,* 1971) suggest that the kinase may be a contaminating enzyme, not functionally related to either of the tubulin subunits. This possibility is strengthened by the finding that vinblastine-induced microtubule crystals contain no kinase activity or cAMP binding activity (Bryan, 1972a). What role (if any) cAMP plays in microtubule function remains a mystery. Such a connection has obvious theoretical attractions, but the existing evidence requires considerable reinforcement.

D. Assembly of Microtubules *in Vitro*

One of the most exciting new areas in microtubule chemistry deals with the *in vitro* assembly of microtubules from soluble tubulin. Initial studies on the reassembly of microtubules were carried out with solubilized outer doublet microtubule proteins from sea urchin sperm tails. Details of this early work have been reviewed by Stephens (1971). In general, this reassembly system required the presence of a divalent cation (Mg^{2+}) and GTP, but was insensitive to temperature or antimitotic agents. Borisy *et al.* (1972) also found that tubulin could aggregate into a number of morphologically different structures, but none of the structures was distinctly tubular. However, this type of aggregation could be prevented by colchicine and low temperature.

The polymerization of tubulin to microtubles indistinguishable from authentic microtubules was first accomplished by Weisenberg (1972). He found that tubulin contained in *crude extracts* of rat brain polymerizes into microtubules upon warming to 37°C in the presence of Mg^{2+}, GTP, and with the removal of Ca^{2+} by prior addition of ethylene-bis(oxyethylene-nitrilo) tetraacetate (EGTA).

Borisy and Olmsted (1972) subsequently reported that the polymerization of porcine brain tubulin is nucleated by disklike structures 290 ± 40 Å in diameter with a 170 ± 20 Å hole, which resemble molecular doughnuts. Removal of the doughnuts by high speed centrifugation (230,000 *g* for 90 minutes) prevents assembly. Addition of a low speed supernatant containing the disklike structures to the high speed supernatant again results in the formation of microtubules. Under these conditions, the assembly of microtubules is temperature dependent and is prevented by colchicine. It is not yet known whether the dislike structures are composed of tubulin, representing intermediates in the assembly of microtubules, or whether they are composed of different proteins. However, they appear to be more stable than microtubules under conditions in which microtubules dissociate to tubulin (low temperature and high Ca^{2+}, or in the presence of colchicine).

We have investigated microtubule assembly in brain supernatant extracts from chick embryos 13 to 15 days old. After homogenization and low speed centrifugation (15,000–25,000 *g* for 20 minutes at 0°C), Mg^{2+}, GTP, and EGTA are added to the supernatants (see Weisenberg, 1972) and the supernatants are warmed at 37°C for 20 minutes. Supernatants undergo gelation, and gelation seems to provide a simple semiquantitative assay for assembly. In this system, gelation (and microtubule assembly) is prevented by colchicine, vinblastine, vincristine, podophyllotoxin; by Cu^{2+}; or by 10^{-3} M Ca^{2+} and low temperature. However, assembly is not prevented by cAMP, Mg^{2+} (10 mM), reducing agents (2-mercaptoethanol or dithiotheritol), or a protease inhibitor (phenyl methyl sulfonyl fluoride). Interestingly, griseofulvin, a substance which produces c-mitotic arrest in plant and mammalian cells, did not appreciably inhibit gelation (see Section IV). Removal of potential nucleating centers by high speed centrifugation (230,000 *g* for 90 minutes) prevents gelation.

Negative staining of the "gel" after fixation in hexylene glycol by the procedure of Weisenberg (1972) or after fixation in glutaraldehyde (Sabatini *et al.*, 1963) reveals that it is composed largely of microtubules. Sectioned pellets of the gel also show classical microtubular elements with 13 protofilaments (J. Bryan and L. G. Tilney, unpublished; L. Wilson, L. Grisham, and K. Bensch, unpublished). The microtubules in griseofulvin-treated extracts appeared identical to those in the control extracts. Incubation of an extract containing *preformed* microtubules with colchicine (10^{-5} *M*) or vinblastine (10^{-5} *M*) for 30 minutes at 37°C did *not* appreciably decrease the number of microtubules as compared with untreated controls. Consequently, as appears to be true for colchicine (Section IV,B), the binding site for vinblastine may be blocked when tubulin is in a structured microtubule. Preliminary experiments indicate that the pellet of reassembled microtubules contains the α and β subunits in equal amounts (J. Bryan, unpublished). The $\alpha:\beta$ ratio was unaffected by a protease inhibitor (phenyl methyl sulfonyl fluoride), which was present during polymerization and subsequent reduction and acetylation.

In vitro assembly systems should provide access to several of the older questions in the literature, such as the role of sulfhydryl groups and the effects of heavy water on mitosis. It also seems reasonable to expect that the temperature dependence, ionic strength, and pH effects and the roles of Ca^{2+} and GTP in microtubule assembly will be understood in the near future. Clearly a large number of questions remain to be answered: What is the role of Ca^{2+} in polymerization? Is there a high-affinity Ca^{2+} binding site on tubulin? Is this site related to the vinblastine binding site (see Wilson *et al.*, 1970)? What is the significance of tubulin phosphorylation? What is the role of guanine nucleotides in the assembly reaction? Is the GTP dephosphorylated during polymerization? What is the chemical nature of the nucleation sites (or doughnuts)? Do these nucleation sites appear at specific times in the cell cycle? A correlated question is whether tubulin is always *competent* to polymerize or whether it changes, perhaps being activated by phosphorylation at a critical time during the cell cycle. Can other cellular organelles serve as *in vitro* nucleation centers, e.g., basal bodies or isolated metaphase chromosomes?

E. Subunit Functions

Considerable indirect evidence (discussed previously) indicates that the colchicine binding receptor is an *oligomer* containing one α and one β subunit. However, almost nothing is known concerning the functions of these two subunits. Different roles could easily be postulated for the subunits during microtubule assembly (e.g., a regulatory role for one subunit; the splitting of GTP for the other). Each subunit may provide unique attachment sites for other macromolecules on the assembled microtubule.

As an initial approach to these questions, Bryan (unpublished) has attempted to determine whether the α subunit, the β subunit or both are involved in the binding of colchicine. Tritium-labeled derivatives of colchicine have been synthesized which potentially can covalently label the active site. Two colchicine derivatives have been synthesized: ring B substituted diazomalonylcolchicine and chlorocyanoethylcolchicine. Both derivatives have been shown to exhibit the same binding specificity for tubulin as chemically unaltered colchicine, and the two equilibrium constants for the binding reaction are similar. Upon irradiation with short wavelength ultraviolet light, the tubulin-bound radioactive photoaffinity labels became covalently attached to a perchloric acid-precipitable, nondialyzable component. Surprisingly, the bulk of the radioactive label could not be detected in either the α or the β tubulin subunits. Rather, the label migrated on SDS polyacrylamide gels as though it were attached to a macromolecule of 16,500 MW. Proteolytic digestions suggested that the component was a polypeptide. It is not clear whether the labeled component is actually an integral part of tubulin, or whether it is a contaminating component that is capable of interacting with colchicine (or the lumicolchicine produced) during photoactivation of the derivatives.

The possibility that there is a third lower molecular weight component of tubulin is an intriguing notion that would explain some puzzling earlier findings. First, in essentially every well analyzed case, the sum of the reported molecular weights of the two subunits has always been conspicuously lower than the molecular weight of the active colchicine-binding receptor. If one adopts the view that the molecular weights of the subunits are equal (52,000), then the molecular weight of a dimer should be 104,000. However, the reported molecular weights of the colchicine–tubulin complex are between 115,000 and 120,000. The 9000 to 16,000 molecular weight difference, perhaps fortuitously, appears to be within the size range for the photoaffinity labeled component. Furthermore, in at least one report on purified tubulin, the appearance of another "component" with a molecular weight of 32,000–33,000 was noted, but not investigated further (Weisenberg *et al.*, 1968).

Heterogeneity within the α and β subunits

Feit *et al.* (1971) and Witman *et al.* (1972a,b) have suggested that each of the two large subunits may themselves be heterogeneous. The basis for this assertion is the finding that subunits from outer doublet microtubules and from brain tubulin can be fractionated further by isoelectric focusing. In this way the outer doublet proteins are separated into five components, while four components can be obtained from neuroblastoma tubulins or immature mouse brain. The basis for this apparent heterogeneity is unclear. It may represent different phosphorylation states of the subunits.

III. Correlation of Ultrastructure and Oligomer Structure

Neither the arrangement of subunits within the active colchicine binding oligomer nor the higher order arrangement of oligomers in the microtubule itself are known with any certainty. Evidence suggesting that each oligomer consists of two large dissimilar subunits (α and β) has already been discussed in Section II. No information exists concerning the intraoligomeric arrangement of subunits.

Because of their marked stability, outer doublet microtubules from flagella and cilia have been studied most thoroughly at a morphological level. Ringo (1967) has proposed a model for outer doublet microtubules, in which there are postulated a total of 23 protofilaments 13 in subfiber A, and 10 in subfiber B. Each protofilament was presumed to be composed of a single row of globular molecules. Three of the A subfiber protofilaments were thought to be shared with the B subfiber. These three protofilaments (termed the "partition") have been isolated (see below) and when negatively stained appear as three adjacent rows of protofilaments, each protofilament being composed of globular molecules.

The usual convention has been to regard each of the globular molecules, as visualized by negative staining, to be a large tubulin subunit (MW 52,000). Recently Warner and Meza (1972) have obtained evidence that challenges the above model. They have isolated tubulin from the B subfiber of sea urchin outer doublet microtubules (see also Wilson and Meza, 1972) and have attempted to resolve the subunit shape and arrangement both within the oligomer and within the protofilaments by negative staining. Their results suggest that the protofilaments within the partition (which in *S. purpuratus* sperm tails contain equimolar amounts of the α and β subunits; Meza *et al.*, 1972; Meza, 1972) are composed of two helically wound subfilaments. Each subfilament has a diameter of 17 ± 2 Å and winds with a half-turn period of 40 Å; it is thought to be the helical arrangement which results in the beaded appearance of the protofilaments. One interpretation would suggest that each unit in a subfilament (which appears to be somewhat S shaped) is a α or β subunit. However, since negative staining of isolated tubulin (oligomers) shows only randomly aggregated 40 Å doughnut-shaped units, the authors were unable to determine whether an oligomer consists of one or two doughnut-shaped particles.

Although the question of which (or how many) subunits are associated with the negative-stained images is of crucial importance for eventually deciphering the substructure of microtubules, several proposed preliminary models deserve discussion; each of these models is based upon a protofilament composed of globular molecules. The common assumption has been that each of the large subunits is a globular molecule with dimensions of approximately 35×40 Å; consequently, an oligomer has been represented as a dimeric structure measuring 35×80 Å. Such globular models agree reasonably well with calculations of the size and shape of the colchicine-binding receptor. The latter calculations have indicated that the oligomer possesses a Stokes radius of 43.5 Å (obtained by replotting gel filtration data in terms of Stokes radii), and an f/f_0 of 1.33 (calculated from the Stokes radius, utilizing a molecular weight of 120,000). Using Perrin's equation (Tanford, 1961), these estimates indicate that the oligomer in solution is a prolate ellipsoid with a/b equal to 5.

Bryan and Wilson (1971) have noted that the constraints of a protofilament and a heterodimer impose rather stringent limitations on possible geometries. An odd number of protofilaments permits only a heteroprotofilament organization in the microtubule. On the other hand, an even number of protofilaments is compatible with either a homo- or heteroprotofilament organization. Unfortunately, the precise number of protofilaments within a microtubule is still not settled, although current evidence suggests that there are 13. X-Ray diffraction studies of microtubules have demonstrated a surface lattice with 40×53 Å packing distances (Cohen *et al.*, 1971); these data are in agreement with an earlier optical diffraction analysis of negatively stained protozoan flagella (Grimstone and Klug, 1966), but are consistent either with 12- or 13-fold rotational symmetry. Some recent evidence obtained with a new staining technique indicates that microtubules in *Echinosphaerium* axonemes and *in vitro* polymerized chick embryo neurotubules contain 13 protofilaments (L. Tilney and J. Bryan, unpublished). The application of a more generalized method of analysis to distinguish the possible packing pattern of subunits in electron micrographs of negatively stained microtubules also strongly suggests that microtubules contain 13 protofilaments (H. P. Erickson, to be published). Thus, cytoplasmic microtubules appear to be constructed of 13 heteroprotofilament strands.

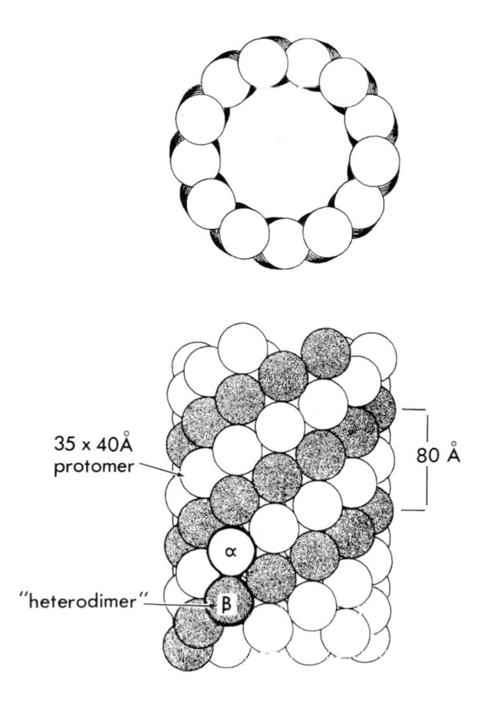

FIG. 3. Microtubule model based upon heteroprotofilament composition.

The microtubule model depicted in Fig. 3 is compatible with the X-ray diffraction data (quasi-hexagonal packing of subunits) and also takes into account the constraints imposed by tubulin chemistry, that is, protofilaments composed of heterodimer units. The microtubule in the model has several interesting features, including *polarity* (which is necessary for the sliding microtubule model of mitosis; Subirana, 1968; McIntosh *et al.*, 1969) and an 80 Å axial repeat, which is necessary to explain the geometrics of various microtubules as found in structures such as cilia and flagella and in *Echinosphaerium* axonemes.

Witman *et al.* (1972a,b) have attempted to define experimentally the substructure of outer doublet microtubules from *Chlamydomonas* flagella, within the context of a globular protofilament model. They sequentially solubilized the doublet microtubules with increasing concentrations of a mild detergent, sodium lauryl sarcosine. They then analyzed the remaining fragments by electron microscopy, and the subunit composition by polyacrylamide gel electrophoresis. Based upon their findings, they proposed a model which has several interesting features: first, the protofilaments are arranged as *homofilaments*. Their data suggest that the homofilaments are paired, i.e., two protofilaments of the same type are adjacent to each other. In addition, their results indicated that the partition region is composed of only a single polypeptide chain; they were able to isolate the "partition," after which polyacrylamide gel electrophoresis of the solubilized structure yielded only a single band, which was called T_1.* These results conflict with those of Meza *et al.* (1972) and of Warner and Meza (1972), who used the outer doublet microtubule partition region of sea urchin sperm tails. Although the S. *purpuratus* sperm tail partition is morphologically similar to the *Chlamydomonas* partition, however, polyacrylamide gel electrophoresis of the sperm tail partition showed equal quantities of both α and β subunits. The reason for this difference between the two systems is not yet known.

Discussion of microtubule morphology and substructure would be incomplete without mention of the other polymorphic forms of tubulin. Vinblastine crystals represent another structural form which will be described in Section IV,C. Although these crystals are composed of tubulin, their colchicine-binding properties are significantly different from those of tubulin solubilized from sea urchin sperm tail outer doublet microtubules. In addition, several morphological studies of microtubules undergoing dissolution suggest that other polymorphic forms may exist. Treatment with antimitotic drugs disrupts microtubules, but this occasionally results in a marked increase of filaments within the cell. Such filaments are usually larger than, and undoubtedly different from, the 50 Å microfilaments found normally. Under certain conditions, protofilaments may be formed. Similarly, dissolution of microtubules by cold (4°–5°C) in some organisms results in the eventual formation of "macrotubules" with a diameter of 360 Å (Tilney and Porter, 1967). Tilney and Porter have proposed that macrotubules are produced by sliding of the protofilaments relative to one another.

IV. Interaction of Drugs with Microtubule Proteins

A. Historical Perspective

Colchicine, an alkaloid (Fig. 4) isolated from the plant *Colchicum autumnale,* has been used for centuries in the treatment of gout. This

* Witman *et al.* (1972a,b) refer to tubulin subunits as Tubulin$_1$ (T_1) and Tubulin$_2$ (T_2). We have utilized the term tubulin to describe the colchichine-binding oligomer; and the Greek letters α, β, to describe each of the subunits. Feit *et al.* (1971) have referred to the subunits as x and y. These nomenclatures are summarized as follows $\alpha = T_1 = x$; $\beta = T_2 = y$.

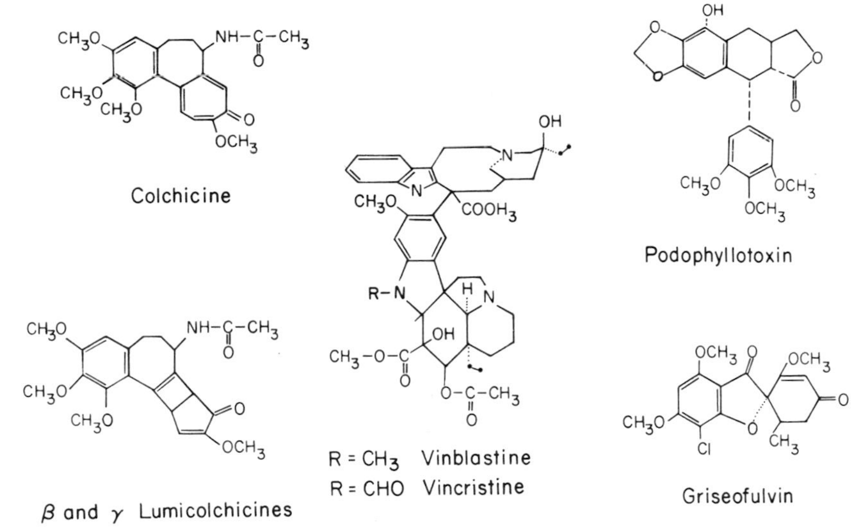

Colchicine

Podophyllotoxin

R = CH_3 Vinblastine
R = CHO Vincristine

Griseofulvin

β and γ Lumicolchicines

FIG. 4. Chemical structures of several antimitotic drugs.

drug was found about 40 years ago to be a potent inhibitor of cell division. In fact, the effect of colchicine on dividing cells was first described by Pernice (1889), but it was not until 1934–1936 that its inhibitory effects on mitosis were recognized (Lits, 1935; Ludford, 1936; Brues, 1936). These early investigations with colchicine which clarified its chemistry, its effects on the mitotic spindle and cellular growth, and its ability to produce polyploidy in plants, were assembled into a comprehensive and classic monograph by Eigsti and Dustin (1955).

The term "c-mitosis" (for colchicine-mitosis) has been introduced to describe the characteristic type of mitotic inhibition produced by spindle poisons, such as colchicine, vinblastine, podophyllotoxin, and griseofulvin (Dustin, 1963). These spindle poisons appeared to produce many similar cytological effects on dividing cells, which implies that they share a common site of action. The recent isolation of receptors for several of these drugs, which turned out to be identical with the protein tubulin, suggests that disruption of microtubule function is their common site of action.

B. COLCHICINE

The earlier investigations with colchicine, as summarized by Eigsti and Dustin (1955), will not be considered here. An important new insight into the mechanism of interaction of colchicine with microtubule protein was provided by Malawista (1965). The various actions of colchicine suggested to him that its effects were related and were due to a decrease in protoplasmic viscosity (a gel-to-sol transformation), possibly caused by the destruction of an organized, labile, fibrillar system concerned with structure and movement within the cell. A methodology necessary for isolating and characterizing this labile fibrillar system was provided by the introduction of tritium-labeled colchicine (Taylor, 1965; Wilson and Friedkin, 1966). Taylor's (1965) early experiments, using cultures of KB cells, demonstrated that labeled colchicine entered the cells and rapidly equilibrated with external colchicine. Moreover, some of the colchicine in the cells assumed a bound form which remained in the cells after removal of colchicine from the external medium.

Taylor demonstrated that low concentrations of colchicine (10^{-7} M) did not affect the rates of DNA, RNA, or protein synthesis, or influence the progress of cells through the cell cycle. Rather, his data were consistent with a mechanism involving the "reversible binding of colchicine to a set of cellular sites," conceivably on the spindle. Furthermore, he suggested that if 3–5% of those sites formed complexes, the cell could not develop a functional mitotic spindle. Borisy and Taylor (1967a) found that much of the colchicine (ring C-methoxy-^{3}H) taken up by cultured KB or HeLa cells was present in the soluble fraction, bound to a macromolecule. Chemically unaltered colchicine was shown to be attached noncovalently. Identical results were obtained by Wilson and Friedkin (1967) with colchicine (acetyl-^{3}H) applied to cultured grasshopper embryos. Similar macromolecular binding of colchicine was obtained *in vitro* in supernatant fractions from HeLa and KB cells, sea urchin eggs, and grasshopper embryos (Borisy and Taylor, 1967a,b; Wilson and Friedkin, 1967).

1. *Evidence for the Specific Binding of Colchicine to the Subunits of Microtubules*

There is now considerable evidence that the high-affinity colchicine receptor in the supernatant fractions of disrupted eukaryotic cells and tissues is a protein subunit of microtubules (tubulin).

Early studies by Borisy and Taylor (1967a) revealed that the relative colchicine binding activity in supernatant extracts of a variety of cells and tissues was approximately correlated with the quantities of microtubules contained in those cells and tissues. Especially high colchicine binding activity was found in brain, which contains very large numbers of microtubules. Macromolecular binding of colchicine in supernatant fractions of grasshopper embryos was also shown to be specifically related to the antimitotic activity of colchicine (Wilson and Friedkin, 1967). For example, a mixture of β and γ lumicolchicines (isomers of colchicine formed by irradiation of colchicine with ultraviolet light; Fig. 4) did not bind to tubulin, nor did high concentrations of the lumicolchicine derivatives inhibit the binding of colchicine to tubulin. This was to be expected, since β and γ lumicolchicines are inactive as c-mitotic agents.

Another approach based on the notion of a common mode of action involved investigating the influence of other c-mitotic agents on the binding of colchicine. The rationale was that other chemical agents which disrupt the mitotic spindle might also affect the binding of colchicine. In fact, podophyllotoxin was found to inhibit colchicine-binding, but vinblastine sulfate actually stimulated colchicine-binding activity. It has since been confirmed that podophyllotoxin binds to tubulin at the colchicine-binding site. The stimulation of colchicine-binding activity by vinblastine sulfate is a consequence of an interaction by vinblastine at a site other than the colchicine-binding site; this interaction stabilizes the protein, a stabilization that gives rise to the apparent stimulation (Section IV,C,2).

With radioactive colchicine as a marker, the colchicine-binding receptor found in the supernatant fraction of porcine brain could be purified to homogeneity by ammonium sulfate fractionation and ion-exchange chromatography on DEAE-Sephadex (Weisenberg *et al.*, 1968). The biochemical and drug-binding properties of the colchicine receptor (tubulin) from brain have been shown to be very similar to those of the proteins solubilized from purified stable microtubules and stabilized labile microtubules. The first direct evidence that the colchicine receptor is derived from microtubules was provided by Shelanski and Taylor (1967), who observed the morphological disappearance of S. *purpuratus* sperm tail central pair microtubules and the concomitant appearance of tubulin in a soluble form. Kirkpatrick *et al.* (1970) isolated assembled microtubules from brain, after stabilization of the labile structures with hexylene glycol; subsequent analysis of the solubilized microtubules revealed that the protein subunits were very similar to those of the colchicine receptor.

More recently, Wilson and Meza (1972) have demonstrated that the colchicine receptor can be recovered from stable sea urchin sperm tail outer doublet microtubules. Previous attempts to demonstrate colchicine binding activity in protein subunits solubilized from outer doublet microtubules were unsuccessful. The reason for this is that the colchicine binding activity of tubulin is unstable and decays in an apparent first-order manner (Weisenberg *et al.*, 1968; Wilson, 1970); and the methods employed previously for the solubilization of the outer doublet proteins, such as dialysis at low ionic strength for 24–48 hours, totally inactivated the protein (see next section).

2. *The Binding of Colchicine to Microtubule Protein: Assaying of Tubulin–Colchicine Complex Formation*

Binding studies with radioactive colchicine and tubulin have provided considerable insight both into the mechanism of action of colchicine and the properties of microtubule proteins. In fact, the colchicine-binding reaction, if employed properly, can be used to quantitate microtubule protein in eukaryotic cells.

Two procedures for the quantitative determination of tubulin–colchicine complexes have been developed: (1) gel filtration; and (2) ionic adsorption of the complex onto disks of DEAE-impregnated filter paper. The first method involves passage of an extract containing tubulin-bound colchicine through a gel filtration column (e.g., a 1×18 cm column of Biogel P-10). The colchicine–tubulin complex separates completely from the free colchicine and can be quantitated accurately (Borisy and Taylor, 1967a,b; Wilson and Friedkin, 1967). The second method involves the use of paper disks that are impregnated with DEAE Sephadex; a number of variations of this separation method exist. The method depends upon the adsorption of the highly acidic tubulin–colchicine complex to the DEAE groups on the paper. Unbound colchicine can be removed either by filtration (Borisy, 1972) or by washing the disks with a low ionic strength buffer (Wilson, 1970). After washing, protein-bound radioactivity attached to the disks can be quantitated by scintillation counting since the colchicine elutes into the organic scintillation liquid. Not only are the paper disk methods rapid and accurate, but also one may process large numbers of samples simultaneously (e.g., 30–40 by the wash procedure). An important disadvantage of the method is that about 20–30% of the colchicine complex breaks down during the procedure, owing to rapid decay of colchicine-binding activity at low ionic strength. Therefore, although binding values are reproducible, they range between 60 and 70% of the values obtained by the gel filtration procedure.

A third method, which has been developed recently (Achor and Wilson, unpublished), promises to be very useful. A 1.5% weight/volume slurry of Dextran T 70 and Norit A (1:5 weight/weight) is prepared. After the colchicine-binding reaction is completed, an equal volume of slurry is added to the incubation mixture, which is then shaken for 10 seconds. Essentially all the unbound colchicine is adsorbed to the Norit A or else trapped in the Dextran particles, while the tubulin-bound colchicine does not adhere to the particles. After centrifugation of the suspension, the free colchicine is pelleted along with the particles; tubulin-bound colchicine may then be quantitated by sampling the remaining supernatant solution.

3. *First-Order Decay of Colchicine-Binding Activity*

Analysis of the colchicine binding reaction has been complicated by the fact that the colchicine-binding activity of tubulin is unstable and decays in an apparent first-order manner (Weisenberg *et al.*, 1968; Wilson, 1970; Bamburg *et al.*, 1973a). Conditions for maintaining optimal stability without the addition of specific stabilizing agents (e.g., vinca alkaloids or GTP) have been worked out for chick embryo brain tubulin (Wilson, 1970; Bamburg *et al.*, 1973a). Figure 5 shows a typical experiment illustrating the apparent first-order decay of colchicine-binding activity by this tubulin under optimal conditions. A 100,000 g supernatant fraction from homogenized chick embryo brain was allowed to age at 37°C in 20 m*M* sodium phosphate–100 m*M* sodium glutamate, pH 6.8 (phosphate–glutamate buffer) at 37°C; every 2 hours, an aliquot was removed for determination of colchicine-binding activity. Since the rate of colchicine complex formation is relatively slow (see Section IV,B,4), incubation with colchicine was carried out for 2 hours. The half-time for decay can be calculated from the slope of the

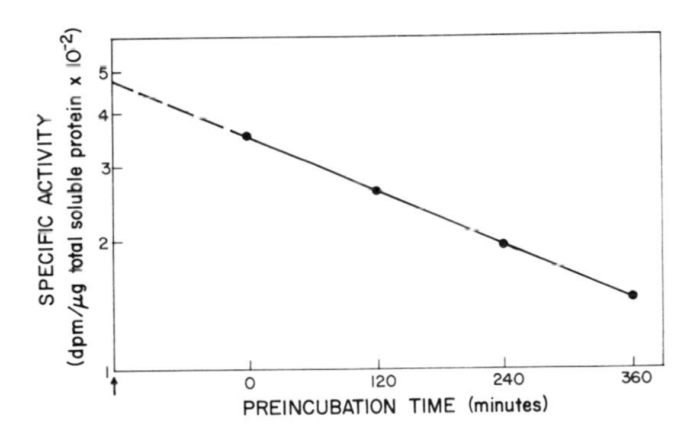

FIG. 5. First-order decay of colchicine-binding activity. A chick embryo brain supernatant fraction was incubated at 37°C. At the times indicated, aliquots were removed and incubated with 2×10^{-6} *M* colchicine (acetyl-^{3}H) for 2 hours at 37°C. Bound colchicine was assayed by the gel filtration method. Extrapolation of the line to include the time of the colchicine incubation (back 120 minutes) yields the *initial binding capacity* of the tubulin. The half-time is calculated from the slope of the line (Wilson, 1970).

line. The initial colchicine-binding activity (i.e., the amount of colchicine that would have been bound if no decay of the tubulin had occurred) can be determined by a 2-hour extrapolation (equal to the time of incubation).

One important parameter that influences the rate of decay is the pH. At 0° and at 37°C, the optimal pH for the maintenance of colchicine-binding activity of chick embryo brain tubulin is 6.7–6.8; stability decreases steeply on each side (Fig. 6). Exposure to a pH of 4.5 or 10.5 for less than 10 seconds irreversibly destroys all colchicine-binding activity. Other important parameters are the ionic strength and the temperature. Colchicine-binding activity of purified chick embryo brain tubulin is most stable at a NaCl concentration of 100 m*M;* the rate of decay increases with lower or higher concentrations of NaCl. With regard to temperature, solubilized sea urchin sperm tail outer doublet microtubule protein (in phosphate-glutamate buffer at pH 6.75) decays with a half-time of 2300 minutes at 0°C; 165 minutes at 37°C; and 5–6 minutes at 50°C. The concentration of active (colchicine-binding) tubulin also affects the rate of decay. Tubulin purified from 13-day-old chick embryo brains incubated at 37°C in phosphate–glutamate buffer pH

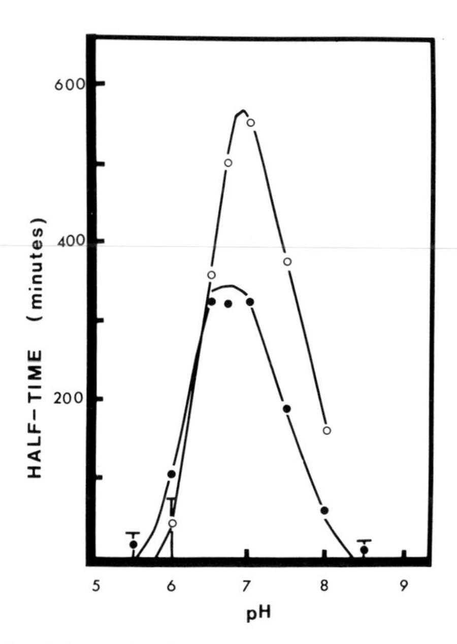

FIG. 6. Stability of colchicine-binding activity as a function of pH (chick embryo brain tubulin); (●——●) 37°C; ○——○, 0°C. (From Wilson, 1970. Reprinted from *Biochemistry* 9, 4999. Copyright 1970 by the American Chemical Society. Reprinted by permission of the copyright owner.)

6.75 decays with a half-time of 120 minutes at a concentration of 20 $\mu g/ml$, and with a half-time of 270 minutes at 240 $\mu g/ml$ (Bamburg *et al.*, 1973a).

The biochemical nature of the inactivation process which tubulin undergoes is not understood. Conceivably the loss of colchicine-binding activity may reflect a change in the protein which is related to the ability of tubulin to polymerize. Guanine nucleotides, which are normally bound to tubulin (see Section II,B) and which appear to be required for the *in vitro* polymerization of microtubules (Weisenberg, 1972; Borisy and Olmsted, 1972), can stabilize the colchicine-binding activity of tubulin (Weisenberg *et al.*, 1968).

4. *The Colchicine-Binding Reaction*

The binding reaction between colchicine and tubulin appears to have the properties of a bimolecular reaction:

$$\text{Colchicine} + \text{tubulin} \rightleftharpoons \text{colchicine–tubulin complex}$$

The binding of colchicine to tubulin is noncovalent; chemically unaltered colchicine can be recovered completely from the complex by extraction with organic solvents or with protein-denaturing agents, such as SDS or urea (Wilson and Friedkin, 1967; Borisy and Taylor, 1967a). The rate of complex formation is slow, especially at low colchicine concentrations. For example, at 2×10^{-6} M the binding of colchicine to chick embryo brain tubulin proceeds to a plateau that is reached after 2–2.5 hours. However, this plateau is reached early due to the decay of colchicine-binding activity.

$$\begin{array}{ccc} \text{Colchicine} + \text{Tubulin} & \underset{k_2}{\overset{k_1}{\rightleftharpoons}} & \text{Colchicine-Tubulin} \\ \downarrow k_3 & & \downarrow k_4 \\ \text{Tubulin}_{(i)} & & \text{Tubulin}_{(i)} + \text{Colchicine} \end{array}$$

SCHEME 1. Inactivation of tubulin. $\text{Tubulin}_{(i)}$ = inactivated tubulin.

In scheme 1, k_1 is the rate constant for association, k_2 is the rate constant for dissociation, and k_3 and k_4 are the rate constants for inactivation of tubulin, which occurs whether or not colchicine is bound. At 37°C k_3 and k_4 are approximately equal (Wilson, 1970). Moreover, k_3 and k_4 are considerably greater than k_2 (to be discussed). Thus, if the protein inactivates at a rate faster than the normal dissociation rate (k_2), the time it takes to reach the plateau (a balance between k_1, k_2, k_3, and k_4) will be considerably less than the time required to reach equilibrium. Supporting this argument is the fact that, after addition of a stabilizing agent, such as vincristine sulfate, at sufficient concentration (1×10^{-5} M) to abolish the decay (minimizing k_3 and k_4), the plateau region is not reached for 7–8 hours (Wilson, unpublished). Furthermore, the binding of colchicine (4.5×10^{-6} M) to tubulin in isolated vinblastine-induced crystals of sea urchin eggs (a system in which colchicine binding activity does not decay) requires approximately 8 hours to reach equilibrium (Bryan, 1972a). This slow rate of association does not seem to be related to the "activation" of the subunit to a form that can bind colchicine, since incubation of the protein at 37°C prior to the addition of colchicine does not significantly increase the binding rate (Wilson and Friedkin, 1967).

A second feature of the colchicine-binding reaction is that the rate depends markedly upon the temperature. This is especially interesting because temperature plays such an important role in the dynamic structure of labile microtubules. At 0° the binding rate is so slow that no binding activity can be detected with tubulin from grasshopper embryos, chick brain, HeLa, or KB cells, even after several hours of incubation (Wilson and Friedkin, 1967; Borisy and Taylor, 1967a,b; Wilson, 1970). However, when the complex is once constituted at 37°C, it is stable at 0°C; thus, the lack of binding at 0°C is not due to an inability to maintain sufficient bonding forces at low temperature. The inability of colchicine to bind to tubulin at low temperatures may explain results obtained many years ago by Hausmann (1906) and by Fühner (1913). Fühner discovered that frogs maintained at 15°–20°C survived doses of colchicine 1000 times larger than they did at 30°C. Similarly, Hausmann observed that bats could survive high doses of colchicine while hibernating in the cold, whereas transferring previously treated bats to warm temperatures resulted in their rapid death. Interestingly, the temperature optimum for the binding of colchicine is not the same for microtubule proteins from all sources. For example, the temperature optimum for tubulin from sea urchin vinblastine-induced crystals appears to be in the region of 20°C (Bryan, 1972a).

A third feature of this binding reaction is that the colchicine binds almost irreversibly. This property of the binding reaction greatly facilitates the quantitation of complex formation by nonequilibrium methods. However, studies on the "tightness" of the colchicine binding reaction have been hampered by the fact that at 37°C inactivation of the protein occurs approximately at the same rate whether or not colchicine is bound. Inactivation of the complex at 37°C results in the release of colchicine, which masks the normal rate of dissociation (Wilson, 1970).

Two experiments can be described that illustrate the tightness of the complex. First, if tubulin containing previously complexed colchicine is stabilized maximally by the addition of vincristine sulfate, then only negligible amounts of colchicine dissociate from the complex within 3 hours at 37°C (Wilson, 1970). Second, after complete saturation of the colchicine-binding sites with unlabeled colchicine, followed by removal of all free colchicine by the Dextran-charcoal procedure (Section IV,B,2), subsequent determinations of the binding rate for labeled colchicine reveal that only 12% of the sites previously occupied by the unlabeled colchicine become available to bind the labeled colchicine after 6 hours of incubation at 37°C.

Some additional information about the nature of complex formation has been obtained by studying the binding reaction with chick embryo brain tubulin at different conditions of pH and ionic strength (Wilson, 1970). Although the inactivation rate of tubulin is affected by variation of the pH between 6 and 8.5, the initial binding activity, and therefore the binding affinity, remains unchanged. Similarly, the rate of inactivation varies considerably with NaCl concentrations between 5 and 500 m*M* without influencing the initial binding capacity (Wilson, 1970). This suggests that the binding forces between colchicine and tubulin are not electrostatic.

Measuring the number of colchicine binding sites on tubulin has been hampered by the lability of the binding sites when the protein is in solution. However, Bryan (1972a) discovered that suspensions of purified vinblastine induced crystals from *S. purpuratus* eggs bind colchicine, and that the colchicine-binding activity of the crystals does not decay. This made it possible to determine the colchicine-binding stoichiometry in the crystal under equilibrium conditions. The value of n (the extrapolated number of binding sites per 110,000 MW of tubulin at infinite colchicine concentration) was 1.07 ± 0.05. This was determined from 12 independent binding experiments at 9 different temperatures.

A value of n has also been obtained for tubulin from outer doublet microtubules of sea urchin sperm flagella, and for tubulin purified from chick embryo brain (L. Wilson, unpublished). In these experiments, incubation of the tubulin with colchicine was carried out for 6–7 hours, and decay was substantially prevented by the addition of vinblastine sulfate. The value of n for outer doublet tubulin was 0.9 ± 0.2, for chick embryo brain tubulin, n was 0.6–0.7 (three determinations). The data suggest that there is one colchicine-binding site on each tubulin molecule (MW 110,000–120,000).

Borisy and Taylor (1967b), using cell-free extracts of *S. purpuratus* eggs and a kinetic analysis, reported a binding constant of 2.3×10^6 liters/mole at 37°C for the colchicine-binding reaction. A similar value of 2.0×10^6 liters/mole has been obtained by us utilizing an equilibrium

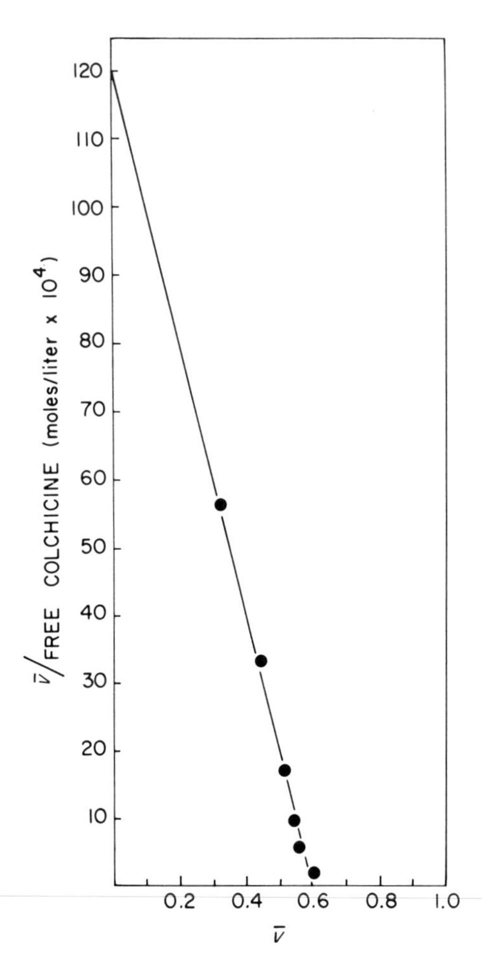

FIG. 7. Scatchard plot: the binding of colchicine to chick embryo brain tubulin. The value of n for this experiment was 0.6. The binding constant (K_A) was 2.0×10^6 liters/mole.

approach with purified chick embryo brain tubulin (Fig. 7). In addition, Bryan (1972a) has determined the binding constant for colchicine at several temperatures, utilizing isolated vinblastine-induced crystals from *S. purpuratus* eggs; Fig. 8 shows the temperature profile for the equilibrium constants of this binding reaction. In such crystals, the binding constants go through an apparent maximum in the region of 20°C. The thermodynamic parameters of the binding reaction could be determined by replotting the data of Fig. 8 in the form of log K vs $1/T$. The values obtained were $\Delta H = 16$ kcal/mole and $\Delta S = 79.5$ eu; at the am-

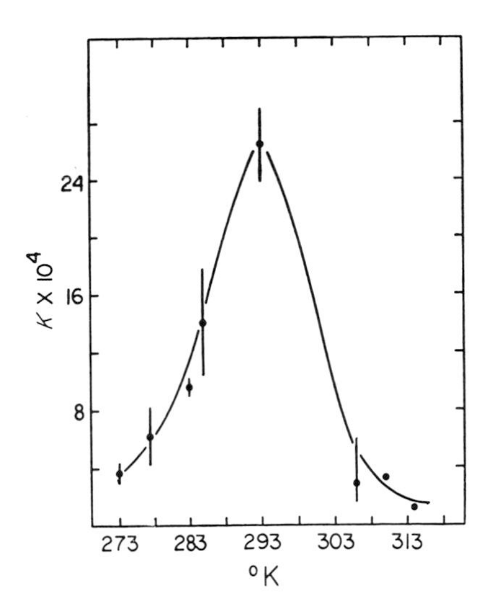

FIG. 8. Equilibrium constants for the colchicine binding reaction at different temperatures; vinblastine crystals from *Strongylocentrotus purpuratus* eggs. (From Bryan, 1972a. Reprinted from *Biochemistry* **11**, 2611. Copyright 1972 by the American Chemical Society. Reprinted by permission of the copyright owner.)

bient temperature of the organism (13°), $\Delta F = -6.7$ kcal/mole. Thus, the binding reaction is characterized by a positive enthalpy, a positive entropy change, and a relatively large, favorable free energy change. These parameters are similar to those measured for the binding of other small apolar molecules, and are consistent with the notion that colchicine binds in a hydrophobic or nonpolar pocket.

5. *The Mechanism of Action of Colchicine*

How does colchicine cause the dissolution of preformed microtubules? If colchicine binds *directly* to the subunits in the microtubule, then (1) the binding site must be available to colchicine when the subunits are assembled; and (2) the interaction of colchicine with its binding site must decrease the binding affinity of at least one of the subunit–subunit association sites, by a conformational change in the protein.

An alternative hypothesis is that colchicine binds only to soluble tubulin, preventing its assembly to microtubules. The rate of dissolution of formed microtubules after addition of colchicine would then be a complex function of the rate of colchicine binding to the soluble tubulin, together with the rates of polymerization and depolymerization of the microtubules. Thus, the more stable the microtubule, the more resistant it would be to the action of colchicine. Support for the latter mechanism has recently been obtained in studies with stable outer doublet microtubules of *S. purpuratus* sperm tails (Wilson and Meza, 1972). It was found that stable microtubules of *S. purpuratus* flagella are not disrupted by colchicine, and the intact microtubules do not bind colchicine. However, once solubilized, the tubulin does possess a high affinity binding site for colchicine.

This colchicine-binding activity of tubulin solubilized from outer doublet microtubules is qualitatively identical to that of other known microtubule proteins. For example, the complex with colchicine has a molecular weight of 115,000, binding activity decays according to first-order kinetics, and decay is prevented by the addition of the vinca alkaloids, vinblastine and vincristine. Lumicolchicines, the photoinactivated isomers of colchicine, do not bind to the tubulin. Finally, this colchicine binding activity is temperature and time dependent, and is competitively inhibited by podophyllotoxin. On the other hand, when this tubulin occurs in intact microtubules, the colchicine-binding site is blocked. After incubation of an outer doublet microtubule suspension with radioactive colchicine, the microtubules can be pelleted by centrifugation and washed; there is no depletion of radioactivity from the remaining supernatant, and no radioactivity is associated with the washed microtubules.

Since it is possible that, upon the binding of colchicine to the structured tubulin, the tubulin–colchicine complex may dissociate from the microtubule, the rate of depolymerization (dissolution) of the microtubules at 37°C in the presence and in the absence of 5×10^{-5} *M* colchicine has been determined. The outer doublet microtubules do dissolve slowly at 37°C, but colchicine does not increase the rate of solubilization. Thus, colchicine does not appear to solubilize outer doublet microtubules, and since the tubulin does possess a colchicine-binding site, that site must be blocked in the assembled microtubule.

6. *The Colchicine-Binding Reaction as an Experimental Tool*

One useful application of the colchicine-binding reaction is for quantitative assay of microtubule protein in eukaryotic cells and tissues. Use of the colchicine-binding reaction for this purpose involves two experimental problems: (1) determination of the amount of colchicine–tubulin complex present in a cell extract at any given time, once that complex has formed (see Section IV,B,2); and (2) determination of the decay rate for loss of colchicine-binding activity.

During the time of incubation used in most investigations (2 hours),

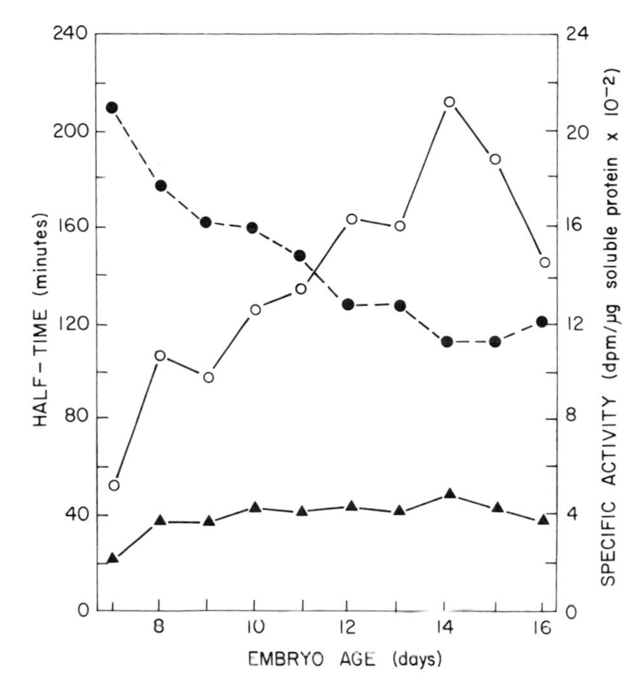

FIG. 9. Comparison of time-decay and single-point colchicine binding assays in supernatants of homogenized ganglia. Colchicine-binding activity in supernatant fractions of homogenized ganglia from different age chick embryos was determined by the time-decay assay procedure. ●---●, Half-time for loss of colchicine binding activity; ○——○, initial binding capacity; ▲——▲, data obtained by use only of the first colchicine binding point of each decay assay (2 hours of incubation with colchicine). Specific activity is expressed as disintegrations per minute of bound colchicine per microgram of total protein in the supernatant fraction. Developmental results obtained in extracts of sonicated and homogenized ganglia differ (Bamburg *et al.*, 1973b).

tubulin progressively loses the ability to bind colchicine; this loss, the rate of which depends upon solution conditions, must be taken into consideration. Thus, the rate of inactivation is measured, and extrapolation of binding data to zero-time of incubation yields a value for the colchicine-binding activity of the tubulin at the beginning of the incubation (Fig. 5). This value, called the *initial binding capacity*, is independent of the rate of decay (for details of the procedure, see Wilson, 1970; Bamburg *et al.*, 1973a).

Figure 9 compares the procedure with a single-point colchicine binding assay in supernatant fractions of *homogenized* chick embryo dorsal root ganglia during development. Results of the single-point assay (2 hours of incubation with colchicine at 37°C) indicate that the concentrations of tubulin in the extracts do not change. However, results with the time-decay assay procedure indicated that there is a 4-fold increase in the tubulin concentration between 6 and 16 days of development. Since the half-time of the binding activity decreases with increasing age, erroneous results are obtained when the single-point assay is used. The tubulin concentrations in these extracts were also determined by an independent method, quantitative polyacrylamide gel electrophoresis on Tris glycinate urea gels (see Bamburg *et al.*, 1973b). The pattern obtained with this technique exactly matched that of the time-decay assay procedure.

The supernatant fractions of embryonic chick brain or ganglia are rich in microtubule protein, and the concentrations can be determined accurately by quantitative polyacrylamide gel electrophoretic procedures. However, gel electrophoresis cannot be utilized to assay microtubule protein in tissues that contain only low concentrations of the protein, since it cannot then be resolved accurately from other proteins present in the gel. Therefore, in these tissue extracts, the time-decay colchicine binding assay procedure must be employed.

Concentrations of microtubule protein in supernatant fractions derived from any tissue source can be *estimated* with the use of the time-decay colchicine-binding procedure in the following way (Bamburg *et al.*, 1973a,b): (1) The concentration of microtubule protein in the supernatant fraction of any tissue rich in the protein (such as chick embryo brain) is determined by the use of quantitative polyacrylamide gel electrophoresis. (2) The initial binding capacity is determined in the same supernatant under standard conditions of pH, ionic strength, temperature, and concentration of radioactive colchicine with known specific activity. (3) The initial binding capacity is then determined in the unknown tissue extract. (4) The microtubule concentration of the unknown extract can then be calculated from the following relationship:

$$\text{Conc. of unknown extract } (\mu\text{g/ml}) = \frac{\text{IBC of unknown extract}}{\text{IBC of known extract}} \times \text{conc. of known extract}$$

where IBC is the *initial binding capacity* given in disintegrations per minute of bound colchicine per microgram of total protein in the extract. This type of analysis requires one assumption: that the affinity of the tubulin in the known tissue is the same as the affinity in the unknown tissue. While this is not strictly true for different species (e.g., the binding constant for chick brain tubulin is 1 to 2×10^6 liters/mole at 37°C, while for sea urchin sperm tail outer doublet tubulin it is 6×10^5 liters/mole; see below and Table IV), it is a reasonable assumption when working with different tissues of single species. For complete accuracy, the binding constant should be determined in the unknown tissue extract.

7. *Biological Effects of Colchicine Unrelated to the Disruption of Microtubules*

A potential hazard associated with the use of the colchicine-binding reaction to measure tubulin quantitatively is the assumption that all macromolecularly bound colchicine in a particular cell extract is bound solely to tubulin. While it is true that few biological effects of colchicine have been found whose mechanisms cannot be traced to an action of colchicine on microtubules, one such action by colchicine is the inhibition of nucleoside transport in cultured mammalian cells (Mizel and Wilson, 1972). This effect of colchicine occurs in HeLa cells, at higher concentrations than those required to inhibit mitosis. Approximately 50% inhibition of uridine, thymidine, adenosine, and guanosine transport occurs at colchicine concentrations of about 4 to 8 $\times 10^{-5}$ *M*, while inhibitory effects on mitosis are seen at concentrations of colchicine as low as 10^{-7} *M*. This effect by colchicine on nucleoside transport appears to be independent of any action of colchicine on microtubules, since (1) β and γ lumicolchicines, derivatives of colchicine which do not inhibit cell division or bind to microtubule protein (Wilson and Friedkin, 1967; Wilson, 1970; Bryan, 1972a), are slightly more potent than colchicine in this system (50% inhibition of thymidine-^{3}H and uridine-^{14}C uptake was obtained with 2 to 5 $\times 10^{-5}$ *M* lumicolchicine); and (2) inhibition of nucleoside uptake by colchicine occurs at 0°C.

These discoveries suggest that still other cellular components may bind colchicine. We have recently found that a macromolecular component of rooster serum (not microtubule protein) also binds colchicine (J. Bamburg and L. Wilson, unpublished). Furthermore, the binding of colchicine in the macromolecular fraction of chick embryo brain is solely to microtubule protein only at colchicine concentrations below 10^{-4} *M*; at higher colchicine concentrations, considerable nonspecific binding occurs (Bamburg *et al.*, 1973a).

Formation of a colchicine–tubulin complex is characterized by a unique set of properties. The presence of at least several of these binding characteristics should be verified when the colchicine-binding reaction is employed as an assay for microtubule proteins (either quantitative or qualitative), especially when it is not possible to purify and characterize the protein. These properties include: (1) temperature dependence; (2) time dependence; (3) first-order decay and stabilization by vinca alkaloids; (4) competitive inhibition by podophyllotoxin; (5) lack of lumicolchicine binding or lack of effect of lumicolchicine on colchicine-binding activity.

C. Vinca Alkaloids

1. *Action of Vinblastine Sulfate and Vincristine Sulfate on Dividing Cells*

The vinca alkaloids are a group of chemically related drugs obtained from the plant *Cantharanthus roseus* G. Don (*vinca rosea* Linn.). Two vinca derivatives, vinblastine and vincristine, have been employed most commonly in studies involving microtubules; both are dimeric indole-dihydroindole derivatives, differing only in the oxidation state of a single carbon atom (Fig. 4). Both vincristine sulfate and vinblastine sulfate are potent mitotic poisons.

The historical development and use of these drugs against certain types of malignant tumors have been described in a recent symposium on vincristine (Sullivan, 1968), as well as in several earlier reports (Cutts *et al.*, 1960; Cutts, 1961; Johnson *et al.*, 1960). The characteristic c-mitotic inhibition produced by vinblastine was first described in mammalian cells by Palmer *et al.* (1960), and similar action was described for vincristine (leurocristine) several years later by Cardinali *et al.* (1963). George *et al.* (1965) demonstrated that c-mitotic inhibition by vincristine sulfate in HeLa cells is accompanied by a dissolution of microtubules; and later Malawista *et al.* (1968) reported a loss of microtubules after spindle disruption by vinblastine sulfate in *Pectinaria gouldi*.

2. *Interaction of Vinca Alkaloids with Microtubules*

The interaction of vinca alkaloids with tubulin in a cell stabilizes the colchicine binding activity of the tubulin (Wilson and Friedkin, 1967) and leads to formation of highly regular tubulin crystals (Bensch and Malawista, 1968, 1969; Schochet *et al.*, 1968; Krishan and Hsu, 1969; Bryan, 1971).

a. Induction of Crystals by Vinca Alkaloids in Vivo. Bensch and Malawista (1969) discovered that the addition of 10^{-5} *M* vincristine sulfate or vinblastine sulfate to L-strain fibroblasts or human leukocytes induced the formation of intracellular crystals within 30 minutes. The size and incidence of these crystals increased with the time of exposure to the alkaloids. After treatment, the cells contained no normal microtubules. Finally, incubation of fibroblasts with 4×10^{-4} M colchicine for 24 hours prior to incubation with vinblastine did not prevent crystal formation.

Sections of crystals parallel to their long axis revealed a highly regular array of electron-opaque lines, composed of regularly spaced dots 240 Å apart (Fig. 10A). Sections perpendicular to the long axis of the crystals showed at low magnification a regular pattern of circles, suggesting that they were cross-sectioned stacks of microtubules (Fig. 10B). At high magnification, the structures appeared to consist of electron lucent centers surrounded by hexagonal rims, which were shared with six surrounding tubules. The rim was 80 Å in width, and the outer diameter of each tubule was 270–280 Å (corresponding to the periodicity of the longitudinally sectioned crystals). On the basis of the striking similarity between microtubules and the microtubule-like structures within the crystal, Bensch and Malawista concluded that the crystals consist mainly of microtubules.

Utilizing an autoradiographic approach with radioactive colchicine, Krishan and Hsu (1971) obtained evidence that vinblastine-induced crystals in HeLa cells possessed colchicine-binding activity, and these investigators also suggested that the crystals contained tubulin. Nagayama and Dales (1970) isolated vinblastine-induced paracrystals from L-strain mouse fibroblasts. Antisera prepared against these paracrystals showed them to share immunological identity with the mitotic apparatus isolated from several types of cells, and they also possessed ATPase activity.

Bryan developed a simple and rapid method for isolation of vinblastine crystals from *S. purpuratus* eggs (Bryan, 1971). These crystals were found to contain equimolar amounts of the α and β subunits of tubulin, while little or no other contaminating protein bands could be detected by urea polyacrylamide gel electrophoresis (Bryan, 1972b). The physical characteristics of the α and β subunits of vinblastine crystal tubulin (Bryan, 1972b) were qualitatively similar to the α and β subunits of chick embryo brain tubulin (Bryan and Wilson, 1971). Another significant finding was that the crystals produced *in vivo* contained 1 mole of vinblastine per mole of tubulin (MW 110,000; Bryan, 1972a). Furthermore, 2 moles of guanosine nucleotide were also bound per mole of tubulin, indicating that the guanosine nucleotide and vinblastine binding sites are different. As discussed previously, the crystals also bound 1 mole of colchicine per mole of tubulin at a third site.

b. Stabilization of Colchicine Binding Activity by Vinca Alkaloids. Spontaneous decay of colchicine-binding activity of tubulin is prevented

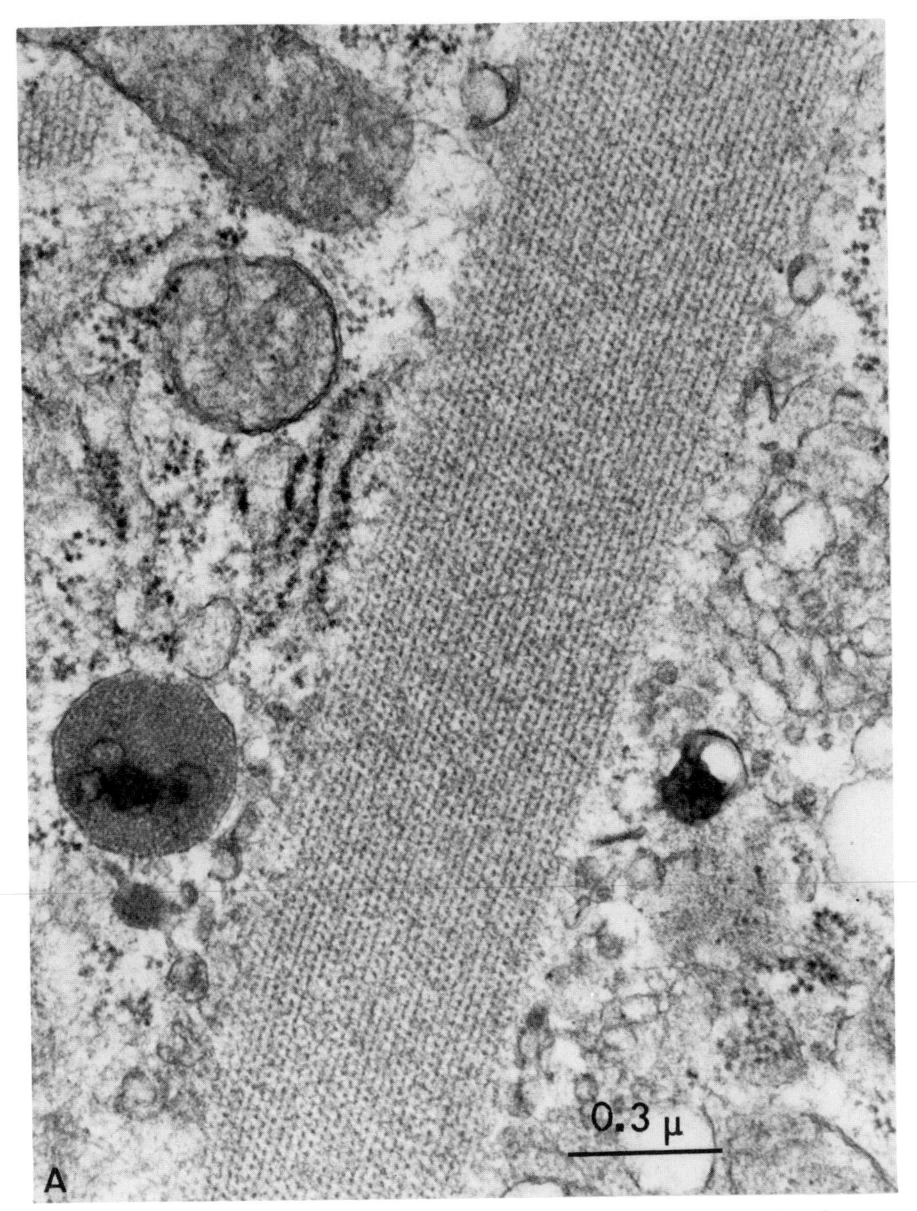

FIG. 10A. Vinblastine crystals. Longitudinal section. (From Bensch and Malawista, 1969.)

by the vinca alkaloids (Wilson and Friedkin, 1967; Wilson, 1970; see Section IV,B,3). At the same time, vinca alkaloids do not affect the affinity between colchicine and tubulin; furthermore, the degree of stabi-

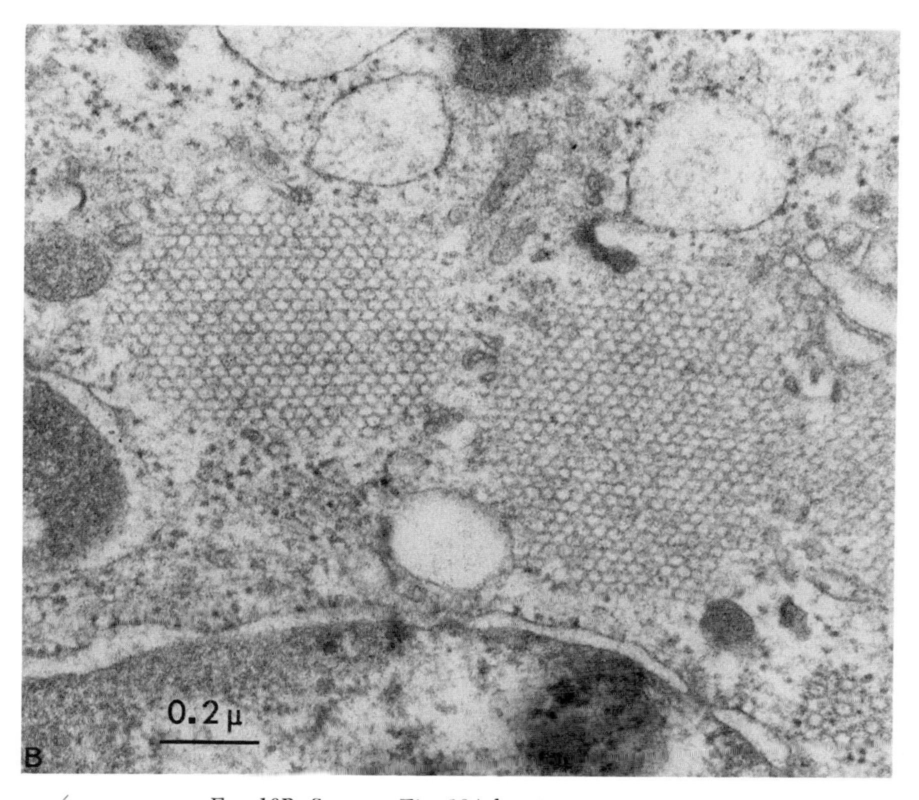

FIG. 10B. Same as Fig. 10A but in cross section.

lization with a given concentration of vinca alkaloid is the same whether or not the tubulin already contains bound colchicine. Table III documents the concentration-dependence for stabilization of colchicine-binding activity by vinblastine sulfate in a chick embryo brain extract at 37°C. Essentially complete stabilization is obtained by the addition of 1 m*M* vincristine sulfate. Also marked stabilization of colchicine-binding activity occurs at vinca alkaloid concentrations well below those necessary to precipitate microtubule protein *in vitro* (see next section).

Ventilla *et al.* (1972) have inferred from circular dichroism spectroscopy that conformational changes do occur in tubulin. They argue that the protein undergoes a slow, temperature-dependent conformational change, resulting in the loss of α helix and β structure at 37°C, and that these changes increase the lability of the protein. Such conformational changes may be related to the first-order loss of colchicine binding activity. It is noteworthy that vinblastine and GTP markedly protected tubulin against the temperature-induced conformational changes. More

TABLE III
EFFECT OF VINBLASTINE AND VINCRISTINE ON THE COLCHICINE-BINDING STABILITY OF TUBULIN AT 37°[a]

Concentration (moles/liter)	Half-time (minutes)	
	Vinblastine	Vincristine
None Control	255	
4.6×10^{-7}	294	410
1.4×10^{-6}	396	600
4.1×10^{-6}	462	1050
1.2×10^{-5}	515	1700
3.7×10^{-5}	790	—
1.1×10^{-4}	965	8450
3.3×10^{-4}	1130	15000
1.0×10^{-3}	2260	~56000

[a] From Wilson (1970). Reprinted from *Biochemistry* **9,** 4999. Copyright 1970 by The American Chemical Society. Reprinted by permission of the copyright owner.

puzzling is the fact that bound colchicine had a similar protective effect on the tubulin, even though colchicine-binding does not appear to prevent spontaneous decay in chick embryo brain tubulin at 37°C. In the experiments of Ventilla *et al.* it is also puzzling that bound colchicine could completely prevent the conformational change, since a considerable portion of the colchicine-binding activity of the tubulin had already decayed.

The ability of a number of vinca alkaloids to stabilize the colchicine-binding activity of tubulin approximately reflects their growth inhibitory activities in cultured Syrian hamster (EHB) cells (Creswell, 1972). The concentration of vinblastine sulfate, vincristine sulfate, or deacetylvinblastine sulfate required to produce mitotic arrest in 50% of the cells was 7.5×10^{-8} *M*. Leurosidine, another dimeric vinca alkaloid, was approximately 100-fold less potent. Stabilization of colchicine binding activity of chick brain tubulin was greatest with vincristine, and slightly less with vinblastine and deacetylvinblastine, while leurosidine was much less potetnt than the other active derivatives tested. Photodegraded vinca alkaloids neither stabilized colchicine-binding activity nor produced c-mitotic effects on the cultured cells. Apparently, the biological activity of the vinca alkaloids resides primarily in the catharanthine moiety (the *upper* indole moiety of the dimeric alkaloid; see Fig. 4), since catharanthine alone was capable of producing mitotic arrest. A mitotic index of 50% was obtained with 5×10^{-5} *M* catharanthine. Vindoline, the *lower* indole moiety of the dimeric alkaloid (Fig. 4), was totally inactive.

Concentrations of vinca alkaloids that stabilize the colchicine binding activity of tubulin also cause the protein to aggregate; this aggregation is characterized by an increase in the sedimentation coefficient from 6 S to approximately 14 S (Weisenberg and Timasheff, 1970). The degree of aggregation depends upon the vinblastine concentration, and begins with vinblastine concentrations below 2×10^{-5} *M*. Similar results have been obtained with vincristine sulfate; addition of 2×10^{-4} *M* vincristine sulfate to a chick embryo brain extract containing 0.1 mg/ml of tubulin increased the sedimentation coefficient of the tubulin from 6.2 to 9.2 S (L. Wilson, unpublished). Weisenberg and Timasheff also measured the protein concentration dependence of vinblastine aggregation; their results extrapolated to a sedimentation coefficient of about 9.5 S at zero protein concentration. Since this value is consistent with a dimer of the 6 S tubulin, the effect of the vinca alkaloids in stabilizing the colchicine-binding activity of tubulin may involve formation of tubulin dimers. In fact, this may be a precursor state for tubulin for crystal formation within a cell.

c. Precipitation of Tubulin by Vinca Alkaloids in Vitro. The demonstration that vinblastine can cause the precipitation of tubulin *in vitro* quickly followed the discovery of crystal formation *in vivo* (Marantz *et al.*, 1969; Bensch *et al.*, 1969; Olmsted *et al.*, 1970; Wilson *et al.*, 1970). Addition of 10^{-3} *M* vinblastine sulfate to a supernatant extract containing tubulin results in precipitation of the tubulin; however, the quantity of tubulin which precipitates depends on the vinblastine concentration, the temperature, and the pH of the buffer. Generally, high concentrations of vinblastine (e.g., 10^{-3} *M*) are required for complete precipitation, but elevated temperatures such as 37°C increase the amount of tubulin which precipitates, and also speed the rate of precipitation. At 0°C precipitation is quantitatively decreased, as is the rate. This temperature dependence of tubulin precipitation by vinblastine may be related to the unusual solubility properties of vinblastine, which itself precipitates at elevated temperatures (Nimni, 1972). Finally tubulin precipitates to the same degree, whether or not it contains bound colchicine.

Several investigators have analyzed the structure of vinblastine-precipitated tubulin by electron microscopy. Bensch *et al.* (1969) found that vinblastine-precipitated porcine brain tubulin contains large areas of ordered structures, which form "ladderlike" arrays. In a similar study, Marantz and Shelanski (1970) argued that vinblastine-induced microtubule crystals are very similar to the structures formed by vinblastine precipitation of tubulin *in vitro*. However, a significant difference between the structures produced *in vitro* and *in vivo* is that *in vivo*-induced vinblastine crystals contain 2 moles of guanine nucleotide per mole of tubulin (Bryan, 1972a). Precipitation of tubulin by vinblastine *in vitro* results in the *release* of tightly bound GTP, and the protection of the exchangeable nucleotide site (Berry and Shelanski, 1972). Consequently, precipitation *in vitro* and crystal formation *in vivo* may proceed by different mechanisms.

Further support for this idea comes from the studies of Wilson *et al.* (1970), who found that vinblastine can precipitate a number of acidic proteins and nucleic acids in addition to tubulin. Those proteins that could be precipitated by vinblastine were also precipitable by calcium ions. These results suggested that vinblastine sulfate, presumably acting as a cation, precipitates proteins by combining with sites that can also combine with Ca^{2+} ions. Addition of vinblastine sulfate to a solution of calf thymus DNA, or to a suspension of polyribosomes, results in complete precipitation of those structures. This probably represents a similar cationlike effect of vinblastine, which may be related to several other actions of vinblastine in cells (see Section IV,C,3).

Vinblastine precipitation of tubulin has been used as a method to purify tubulin from crude supernatant extracts of several tissues (Olmsted *et al.*, 1970; Dutton and Barondes, 1969). Although vinblastine precipitation is evidently a valuable procedure for purification of tubulin, it is not as highly selective as was once thought. The relatively high degree of success with this procedure can be attributed to the use of cell extracts containing high concentrations of tubulin, as compared with low concentrations of other precipitable proteins. Vinblastine precipitability by itself clearly cannot be employed for the specific identification of tubulin in cell extracts. We have been unable to precipitate tubulin with vincristine sulfate (Wilson *et al.*, 1970), even though vincristine induces crystal formation *in vivo*. This is another difference between *in vitro* and *in vivo* produced tubulin structures which argues for the involvement of different mechanisms.

d. Possible Mechanisms of Action of the Vinca Alkaloids. Conceivably there may be two different types of binding sites for vinblastine on tubulin. Interaction of vinblastine with a high affinity site (or sites), which stabilizes the colchicine-binding activity and causes the aggregation of tubulin, would result in the disruption of microtubules and crystal formation within the cell. On the other hand, interaction of vinblastine with perhaps a large number of low affinity sites would result in the precipitation of tubulin *in vitro*.

It is not known whether vinblastine can directly disrupt a preformed microtubule. If it cannot, as appears to be the case with colchicine, a possible mechanism of microtubule disruption would involve the binding of vinblastine to a high affinity site on soluble tubulin. This might prevent a conformational change necessary for microtubule assembly. Aggregation of tubulin into tubulin dimers would be the initial step in the formation of crystals within a cell. Since the crystals contain 1 mole of vinblastine per mole of tubulin, it is unlikely that crystal formation involves cross bridging of two tubulin molecules with one vinblastine molecule. If the colchicine-binding site is one of the association sites for normally assembled microtubules, as the evidence now suggests, we can argue that this site is not involved in the assembly of tubulin in a vinblastine crystal (Bryan, 1972a). On the other hand, at least two protein association sites would be predicted in a normal microtubule, since there are both lateral and vertical attachment sites; only one of these could correspond to the colchicine-binding site. Consequently aggregation of tubulin dimers to form crystals could proceed by association at one or several of the other association sites of tubulin.

3. *Other Actions of the Vinca Alkaloids*

The effects of the vinca alkaloids on mammalian cells clearly are much more complex than the foregoing sections would indicate. These compounds do not share the property of colchicine, which at low concentrations appears to interact with only a few receptors within the cell. Rather, the vinca alkaloids seem to affect a variety of unrelated cellular processes.

For example, vinca alkaloids have been shown to inhibit the incorporation of uridine-^{3}H into RNA (Wagner and Roizman, 1968; Augustin and Creasey, 1967). This effect may be due to the inhibition of uridine-^{3}H uptake into the cell, as well as to specific inhibition of RNA synthesis (Plagemann, 1972). Krishan and Hsu (1969) have also observed arrays of helical polyribosomes in L cells. Similarly, aggregation of ribosomes has been observed in a prokaryotic cell, *Escherichia coli*, which is thought to contain no microtubules (Kingsbury and Voelz, 1969). Perhaps an aggregation of ribosomes similar to that occurring *in vitro* (Wilson *et al.*, 1970) also occurs in cells, brought about through the cationlike activity of vinblastine. Clearly, much more research is warranted in this area.

D. Podophyllin Alkaloids: Podophyllotoxin

Podophyllotoxin (Fig. 4) is one of the principal alkaloids of podophyllum resin; it is obtained by extraction of the dried rhizome and roots of *Podophyllum peltatum* L. with ethanol (Kelly and Hartwell, 1954).

King and Sullivan (1946, 1947) first recognized the similarity between the actions of podophyllotoxin and colchicine on dividing cells. In general, the antimitotic activity of podophyllotoxin is qualitatively indistinguishable from that of colchicine. Early investigations on the biological properties of the podophyllum alkaloids have been described in several reviews (Cornman and Cornman, 1951; Kelly and Hartwell, 1954; von Wartburg *et al.*, 1957; Seidlova-Masinova *et al.*, 1957).

1. *Interaction of Podophyllotoxin with Tubulin*

The first biochemical evidence that the mechanism of action of podophyllotoxin is very similar to that of colchicine was the finding that podophyllotoxin prevents the binding of colchicine to grasshopper embryo tubulin (Wilson and Friedkin, 1967). Picropodophyllotoxin, an isomer of podophyllotoxin exhibiting considerably weaker antimitotic activity, was less potent than podophyllotoxin in its ability to inhibit the binding of colchicine both in living embryos and in cell-free extracts.

That podophyllotoxin binds strongly to tubulin at the colchicine-binding site is now firmly established. The ability of podophyllotoxin to prevent colchicine binding in supernatant fractions of chick embryo brain is concentration dependent; furthermore, a preformed colchicine–tubulin complex is not destroyed by addition of podophyllotoxin. These facts suggest that podophyllotoxin does not inhibit the binding of colchicine by an allosteric mechanism (Wilson, 1970). More directly, Bryan (1972a) has found that podophyllotoxin inhibits the binding of colchicine to tubulin in sea urchin egg vinblastine crystals by a competitive mechanism. Similar results have been obtained with purified chick embryo brain tubulin (Fig. 11) and with solubilized outer doublet tubulin from *S. purpuratus* sperm tails (Wilson and Meza, 1972).

The inhibition constants for podophyllotoxin relative to that of colchicine for tubulins from several sources, and at 37°C appear in Table IV. The inhibition constant (K_i) is analogous to the dissociation constant (K_D), which describes the affinity of the binding site for the ligand. The affinity of brain tubulin for podophyllotoxin is approximately twice that for colchicine. Similarly, although the affinity of outer doublet tubulin for colchicine is reduced by 2- to 3-fold as compared with brain tubulin, the relative affinities for colchicine and podophyllotoxin are essentially unchanged. An interesting difference appears in the case of vinblastine crystals: here the affinity for colchicine is approximately ten times lower than that of brain tubulin, but the affinity for podophyllotoxin is decreased almost 75 times. Consequently, the relative affinities for the two drugs are reversed for tubulin in crystals. Whether the altered affinities represent real differences between the tubulin of brain

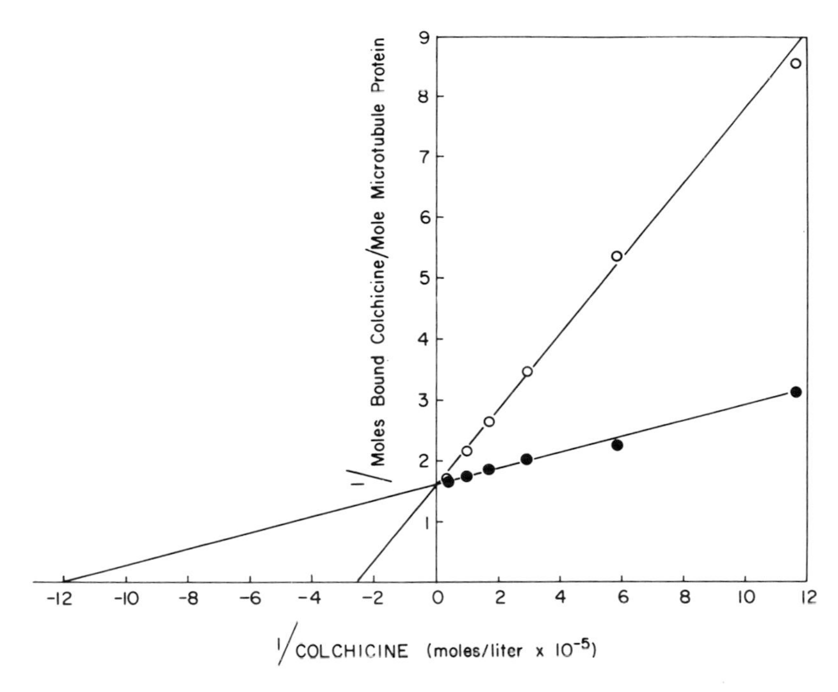

FIG. 11. Competitive inhibition of colchicine binding activity by podophyllotoxin. Purified chick embryo brain tubulin. ●——●, Colchicine alone; ○——○, plus 2×10^{-6} *M* podophyllotoxin.

TABLE IV
RELATIVE AFFINITIES OF TUBULIN FOR COLCHICINE AND PODOPHYLLOTOXIN AT 37°C

Microtubule source	Colchicine K_D ($\times 10^{-7}$)	Podophyllotoxin $(K_i) K_D$ ($\times 10^{-7}$)	K_D colchicine / K_D podophyllotoxin
Chick embryo brain (purified)	9.1	3.9	2.3
Chick embryo brain (crude extract)	8.5	5.4	1.6
Outer-doublet microtubules (*Strongylocentrotus purpuratus* sperm tails)	21	13	1.6
Vinblastine crystals (*S. purpuratus* eggs)[a]	85	300	0.28

[a] At 13°C.

and that contained in the vinblastine crystals, or whether the differences are due to the arrangement of tubulin in the crystal itself, remains to be determined. The inhibition constant for picropodophyllotoxin with purified brain tubulin is 2.7×10^{-5}, almost 100 times weaker than for podophyllotoxin; this is in keeping with its decreased biological activity.

Several features of the binding reaction between podophyllotoxin and tubulin are notably different from that between colchicine and tubulin. For example, the podophyllotoxin-tubulin binding reaction is complete at 37°C in less than 10 minutes, it occurs readily at 0°, and it is more rapidly reversible than is the binding of colchicine. These experiments were done using colchicine binding activity as an assay; a more complete study will be possible when tritium-labeled podophyllotoxin becomes available.

2. *Other Biological Effects of Podophyllotoxin*

Since podophyllotoxin and colchicine have similar sites of action, podophyllotoxin should disrupt any microtubule system which is disrupted by colchicine; this appears to be the case. Although there are only a few documented observations at the electron microscope level demonstrating microtubule disruption by podophyllotoxin (e.g., Wisniewski *et al.*, 1968), a number of reports indicate that those processes which are dependent upon microtubules and are altered or inhibited by colchicine are similarly affected by podophyllotoxin (Williams and Wolff, 1970; Taylor *et al.*, 1973; Makrides *et al.*, 1970).

In addition, as with colchicine, podophyllotoxin has actions on cells which appear to be unrelated to any effect on microtubules. For example, podophyllotoxin competitively inhibits nucleoside transport in mammalian cells. The ability of podophyllotoxin to inhibit nucleoside transport is approximately 20 times greater than that of colchicine (K_i for colchicine = 6×10^{-5} *M*; for podophyllotoxin, 3×10^{-6} *M*). Here, picropodophyllotoxin is only slightly less potent than podophyllotoxin (K_i for picropodophyllotoxin = 7×10^{-6} *M*; Mizel and Wilson, 1972, and unpublished data). Moreover, a number of investigators have described inhibitory effects on several enzyme systems (Kelly and Hartwell, 1954). Consequently, as with colchicine, any attempt to relate an action of podophyllotoxin to an interaction with microtubules requires additional supportive evidence.

E. Miscellaneous c-Mitotic Chemical Agents

1. *Griseofulvin*

Griseofulvin (Fig. 4) is a mold metabolite isolated from several species of mold belonging to the genus *Penicillium*, e.g., *Penicillium griseofulvin.*

This drug has been used clinically for the treatment of specific dermatophyte infections and was found to inhibit cell division in a manner similar to that of colchicine in rat bone marrow and root tips of *Vicia faba* (Paget and Walpole, 1958, 1960). Deysson studied the action of griseofulvin on the meristem cells of *Allium cepa* (Deysson, 1964a) and of cultured HeLa cells (Deysson, 1964b). The c-mitotic figures produced were indistinguishable from those produced by colchicine. Malawista *et al.* (1968) reported that the ability of 10^{-5} *M* griseofulvin to destroy a preformed mitotic spindle was remarkably rapid in oocytes of *Pectinaria gouldi*, requiring only about 5 minutes. Moreover, the effect was completely and rapidly reversible, with full recovery 5.5–11 minutes after removal of the drug. In addition, the regeneration of cilia in *Stentor* is blocked by griseofulvin, as it is by other chemicals that can disrupt microtubule function (Margulis *et al.*, 1969).

However, a griseofulvin concentration that inhibits mitosis almost completely in grasshopper embryos or in HeLa cells (6×10^{-5} *M*) has no effect on the binding of tritium-labeled colchicine to grasshopper embryo or chick embryo brain tubulin (Wilson, 1970). If it were true that colchicine and griseofulvin share the same binding site on tubulin, then (as in the case of podophyllotoxin) griseofulvin should prevent the binding of colchicine. Neither does griseofulvin influence the rate of decay of colchicine binding activity, a process that is largely prevented by the vinca alkaloids and by GTP. Thus, if griseofulvin does interact directly with tubulin, its binding site is likely to be different from that of the vinca alkaloids or guanine nucleotides.

As described earlier, it has recently become possible to polymerize microtubules *in vitro* in crude supernatant fractions of brain which contain tubulin by the removal of calcium ions by chelation with ethylenebis-(oxyethylene-nitrilo) tetraacetate (EGTA) (Weisenberg, 1972; Borisy and Olmsted, 1972). We have investigated the effect of griseofulvin on this assembly reaction, and have found that griseofulvin does not prevent microtubule polymerization *in vitro* as assayed by electron microscopy. Moreover, examination of griseofulvin-inhibited HeLa cells by electron microscopy revealed the presence of morphologically normal microtubules (Fig. 12) (Grisham *et al.*, 1973). Thus the action of griseofulvin on mitosis, though resembling in a number of ways that of other c-mitotic agents such as colchicine, podophyllotoxin, and the vinca alkaloids, appears to differ from these agents in that microtubule assembly may not be affected.

How does griseofulvin inhibit mitosis? One possibility is that it affects a process vital to the sliding of microtubules, which has been proposed to be necessary for separation of chromosomes (McIntosh *et al.*, 1969).

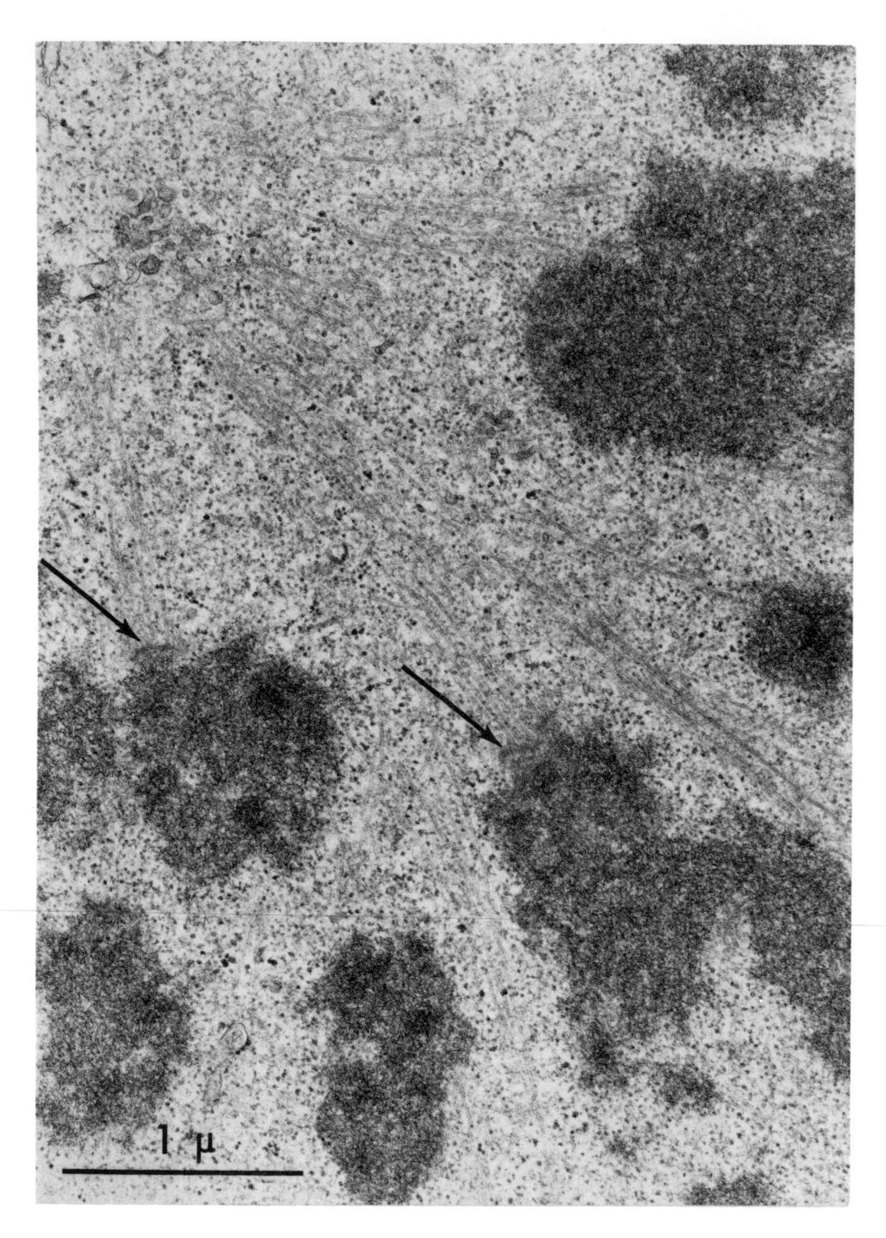

FIG. 12. Section of a HeLa cell blocked in mitosis by exposure to 2×10^{-5} *M* griseofulvin for 24 hours. Note the normal appearance of the microtubules and normal attachment to kinetochores.

Alternatively griseofulvin may not directly affect microtubules at all, but may act on other processes essential for normal completion of cell division. Further understanding of the mechanism of action of this drug on mitosis may provide new insight into the process of cell division of eukaryotic cells.

2. *Heavy Metals, Sulfhydryl Reagents, Carbamates, General Anesthetics*

In addition to drugs described in previous sections, a variety of diverse chemical agents are thought to exert disruptive effects on microtubule function. Among these are the general anesthetic gases, chloral hydrate (Andersen, 1966; Kennedy and Brittingham, 1968; Allison and Nunn, 1968; Allison *et al.*, 1970), sulfhydryl reagents (Mazia and Zeuthen, 1966; Wade and Satir, 1968), and heavy metals (Roth and Shigenaka, 1970). Colchicine binding is not influenced by chloral hydrate, by sulfhydryl reagents such as mercaptoethanol and dithioerythritol, by ether, or by isopropyl-*N*-phenylcarbamate (L. Wilson, unpublished); nevertheless, this finding does not preclude the possibility that such agents interact with microtubule proteins.

Several heavy metals, including copper and nickel, disrupt the axopods of *Echinosphaerium nucleofilium,* and their actions appear to be mediated via the disruption of microtubules. We have investigated the interaction of copper(II) with the tubulin of chick embryo brain because (1) copper forms complexes with guanine nucleotides (Tu and Friederich, 1968) and the action of copper on microtubules may be mediated via an effect on the tubulin-bound GTP; and (2) the interaction between copper and microtubules may be related to the neurological damage characteristic of Wilson's disease, which involves accumulation of high copper concentrations in the central nervous system. One result is that copper inhibits the binding of colchicine, with 50% inhibition at 2×10^{-5} *M*. However, this apparent inhibition of colchicine-binding activity by copper(II) is caused by a marked increase in the rate of decay of colchicine-binding activity; thus, inhibition of colchicine-binding activity is not necessarily due to the binding of the inhibitor to the colchicine-binding site. The concentration range for inactivation of tubulin *in vivo* is similar to that which results in disruption of microtubules in *Echinosphaerium.* Interestingly, nickel ions (which also disrupt microtubules in *Echinosphaerium*) do not influence the colchicine-binding activity of tubulin; therefore, the actions of nickel and copper on microtubules must be different.

Another interesting drug effect on microtubules is that caused by isopropyl-*N*-phenylcarbamate (IPC) on plant cell mitosis. Disorganization of the mitotic spindle produced by IPC is not due to the dissolution of microtubules, but to the disorientation of intact microtubules (Hepler and Jackson, 1969) in a manner somewhat similar to the action of griseofulvin. This appears to be mediated via an action on the poles of the spindle. It is important to emphasize that either dissolution or disorientation of microtubules by chemical agents may be produced *indirectly* by an action on critical cellular components, which are required for the structural integrity of microtubules.

V. Summary

Much evidence indicates that the 120,000 molecular weight colchicine receptor found in the soluble fraction of eukaryotic cells is the building block of microtubules (tubulin). It is a dimer composed of two subunits. Each of these subunits is different, and therefore probably serves different functions during or after microtubule assembly. The list of ligands interacting with tubulin includes antimitotic drugs, calcium, and guanine nucleotides; this list could easily be increased to include phosphate (phosphorylation or transphosphorylation activities) and macromolecules possessing ATPase (or GTPase) activities, e.g., dynein. The mechanism for assembly of the oligomers into microtubules, and the inhibition of the assembly process by antimitotic agents, remain interesting and exciting areas for research. Recent advances demonstrating self-assembly of microtubules *in vitro* should provide a tremendous impetus to research in this area. Some of the questions now open to investigation have been discussed.

The arrangement of tubulin subunits within the oligomer, and of the oligomer within the microtubule, remain open areas that deserve considerable attention. This general problem becomes more fascinating with the possibility of studying the various polymorphic forms of tubulin (microtubules, macrotubules, and protofilaments), together with the interactions between microtubules and other macromolecules, such as dynein, which may play an important role in microtubule function.

Acknowledgments

These investigations have been supported by USPHS grant No. NS09335 from the National Institute of Neurological Diseases and Stroke; American Cancer Society Grant No. E603 to Dr. L. Wilson; and grant No. GB32287X from the National Science Foundation to Dr. J. Bryan. The literature search for this review was completed in December, 1972.

REFERENCES*

Allison, A. C., and Nunn, J. F. (1968). *Lancet* **II**: 1326.
Allison, A. C., Hulands, G. H., Nunn, J. F., Kitching, J. A., and MacDonald, A. C. (1970). *J. Cell Sci.* **7**: 483.
Andersen, N. B. (1966). *Acta Anaesthesiol. Scand., Suppl.* **22**: 3.
Augustin, B. M., and Creasey, W. A. (1967). *Nature (London)* **215**: 965.
Bamburg, J. R., Shooter, E. M., and Wilson, L. (1973a). *Biochemistry* **12**: 1476.
Bamburg, J. R., Shooter, E. M., and Wilson, L. (1973b). *Neurobiology* 3: 162.
Behnke, O., and Forer, A. (1967). *J. Cell Sci.* **2**: 169.
Bensch, K. G., and Malawista, S. E. (1968). *Nature (London)* **218**: 1176.
Bensch, K. G., and Malawista, S. E. (1969). *J. Cell Biol.* **40**: 95.
Bensch, K. G., Marantz, R., Wisniewski, H., and Shelanski, M. (1969). *Science* **165**: 495.
Berry, R. W. and Shelanski, M. L. (1972). *J. Mol. Biol.* **71**: 71.
Bibring, T., and Baxandall, J. (1971). *J. Cell Biol.* **48**: 324.
Borisy, G. G. (1972). *Anal. Biochem.* **50**: 373.
Borisy, G. G., and Olmsted, J. B. (1972). *Science* **177**: 1196.
Borisy, G. G., and Taylor, E. W. (1967a). *J. Cell Biol.* **34**: 525.
Borisy, G. G., and Taylor, E. W. (1967b). *J. Cell Biol.* **34**: 535.
Borisy, G. G., Olmsted, J. B., and Klugman, R. A. (1972). *Proc. Nat. Acad. Sci. U.S.* **69**: 2890.
Brues, A. M. (1936). *J. Physiol. (London)* **86**: 63P.
Bryan, J. (1971). *Exp. Cell Res.* **66**: 129.
Bryan, J. (1972a). *Biochemistry* **11**: 2611.
Bryan, J. (1972b). *J. Mol. Biol.* **66**: 157.
Bryan, J., and Wilson, L. (1971). *Proc. Nat. Acad. Sci. U.S.* **68**: 1762.
Cardinali, G., Cardinali, G., and Enein, M. A. (1963). *Blood* **21**: 102.
Cohen, C., Harrison, S. C., and Stephens, R. E. (1971). *J. Mol. Biol.* **59**: 375.
Cornman, I., and Cornman, M. E. (1951). *Ann. N.Y. Acad. Sci.* **51**: 1443.
Creswell, K. (1972). Master's Thesis, Stanford Univ., Stanford, California.
Cutts, J. H. (1961). *Cancer Res.* **21**: 168.
Cutts, J. H., Beer, C. T., and Noble, R. L. (1960). *Cancer Res.* **20**: 1023.
Deysson, G. (1964a). *Ann. Pharm. Fr.* **22**: 17.
Deysson, G. (1964b). *Ann. Pharm. Fr.* **22**: 89.
DiBella, F. P., Goodman, D. B. P., and Rasmussen, M. (1971). *Fed. Proc., Fed. Amer. Soc. Exp. Biol.* **30**: 1194.
Dustin, P., Jr. (1963). *Pharmacol. Rev.* **15**: 449.
Dutton, G. R., and Barondes, S. (1969). *Science* **166**: 1637.
Eigsti, O. J., and Dustin, P., Jr. (1955). "Colchicine in Agriculture, Medicine, Biology and Chemistry." Iowa State Coll. Press, Ames, Iowa.
Eipper, B. A. (1972). *Proc. Nat. Acad. Sci. U.S.* **69**: 2283.
Everhart, L. P. (1971). *J. Mol. Biol.* **61**: 745.
Falxa, M. L., and Gill, T. J., III (1969). *Arch. Biochem. Biophys.* **135**: 194.
Feit, H., Slusarek, L., and Shelanski, M. L. (1971). *Proc. Nat. Acad. Sci. U.S.* **68**: 2028.
Fine, R. E. (1971). *Nature (London), New Biol.* **233**: 283.
Führer, H. (1913). *Arch. Exp. Pathol. Pharmakol.* **72**: 228.

* The literature search for this review was completed in December, 1972.

George, P., Journey, L. J., and Goldstein, M. N. (1965). *J. Nat. Cancer Inst.* **35**: 355.
Goodman, D. B. P., Rasmussen, H., DiBella, F., and Guthrow, C. E. (1970). *Proc. Nat. Acad. Sci. U.S.* **67**: 652.
Grimstone, A. V., and Klug, A. (1966). *J. Cell Sci.* **1**: 351.
Grisham, L., Wilson, L., and Bensch, K. G. (1973). *Nature (London)* **244**: 294.
Hausmann, W. (1906). *Arch. Gesamte Physiol. Menschen Tiere* **113**: 317.
Hepler, P. K., and Jackson, W. T. (1969). *J. Cell Sci.* **5**: 727.
Inoué, S., and Sato, H. (1967). *J. Gen. Physiol.* **50,** Suppl.: 259.
Jacobs, M., and McVittie, A. (1970). *Exp. Cell Res.* **63**: 53.
Johnson, I. S., Wright, H. F., Svoboda, G. H., and Valentis, J. (1960). *Cancer Res.* **20**: 1017.
Kelly, M. G., and Hartwell, J. L. (1954). *J. Nat. Cancer Inst.* **14**: 967.
Kennedy, J. R., Jr., and Brittingham, E. (1968). *J. Ultrastruct. Res.* **22**: 530.
King, L. S., and Sullivan, M. (1946). *Science* **104**: 244.
King, L. S., and Sullivan, M. (1947). *Arch. Pathol.* **43**: 374.
Kingsbury, E. W., and Voelz, H. (1969). *Science* **166**: 768.
Kirkpatrick, J. B., Hyams, L., Thomas, V. L., and Howley, P. M. (1970). *J. Cell Biol.* **47**: 384.
Krishan, A., and Hsu, D. (1969). *J. Cell Biol.* **43**: 553.
Krishan, A., and Hsu, D. (1971). *J. Cell Biol.* **48**: 407.
Lagnado, J. R., Lyons, C., Weller, M., and Phillipson, O. (1972). *Biochem. J.* **128**: 95P.
Lits, F. J. (1935). *C. R. Soc. Biol.* **118**: 393.
Ludford, R. J. (1936). *Arch. Exp. Zellforsch. Besonders* **18**: 411.
Luduena, R. F., and Woodward, D. O. (1972). *Fed. Proc., Fed. Amer. Soc. Exp. Biol.* **31**: 882.
McIntosh, J. R., Hepler, P. K., and Van Wie, D. G. (1969). *Nature (London)* **224**: 659.
Makrides, E. B., Banerjee, S., Handler, L., and Margulis, L. (1970). *J. Protozool.* **17**: 548.
Malawista, S. E. (1965). *J. Exp. Med.* **122**: 361.
Malawista, S. E., Sato, H., and Bensch, K. G. (1968). *Science* **160**: 770.
Marantz, R., and Shelanski, M. L. (1970). *J. Cell Biol.* **44**: 234.
Marantz, R., Ventilla, M., and Shelanski, M. (1969). *Science* **165**: 498.
Margolis, R. K., Margolis, R. U., and Shelanski, M. L. (1972). *Biochem. Biophys. Res. Commun.* **47**: 432.
Margulis, L., Banerjee, S., and Neviakas, J. A. (1969). *J. Protozool.* **16**: 660.
Mazia, D., and Zeuthen, E. (1966). *C. R. Trav. Lab. Carlsberg* **35**: 341.
Meza, I. (1972). Ph.D. Thesis, Univ. of California, Berkeley, California.
Meza, I., Huang, B., and Bryan, J. (1972). *Exp. Cell Res.* **74**: 535.
Miyamoto, E., Kuo, J. F., and Greengard, P. (1969). *J. Biol. Chem.* **244**: 6395.
Mizel, S. B., and Wilson, L. (1972). *Biochemistry* **11**: 2573.
Nagayama, A., and Dales, S. (1970). *Proc. Nat. Acad. Sci. U.S.* **66**: 464.
Nimni, M. E. (1972). *Biochem. Pharmacol.* **21**: 485.
Olmsted, J. B., and Borisy, G. G. (1973). *Annu. Rev. Biochem.* **42**: 507.
Olmsted, J. B., Carlson, K., Klebe, R., Ruddle, F., and Rosenbaum, J. (1970). *Proc. Nat. Acad. Sci. U.S.* **65**: 129.
Owellen, R. J., Owens, A. H., Jr., and Donigian, D. W. (1972). *Biochem. Biophys. Res. Commun.* **47**: 685.

Paget, G. E., and Walpole, A. L. (1958). *Nature (London)* **182**: 1320.
Paget, G. E., and Walpole, A. L. (1960). *AMA Arch. Dermatol.* **81**: 750.
Palmer, C. G., Livengood, D., Warren, A. K., Simpson, R. J., and Johnson, I. S. (1960). *Exp. Cell Res.* **20**: 198.
Pernice, B. (1889). *Sicil. Med.* **1**: 265.
Plagemann, P. G. (1972). *J. Nat. Cancer Inst.* **45**: 589.
Porter, K. R. (1966). *Principles Biomol. Organ., Ciba Found. Symp. 1965* p. 308.
Renaud, F. L., Rowe, A. J., and Gibbons, I. R. (1968). *J. Cell Biol.* **36**: 79.
Ringo, D. L. (1967). *J. Ultrastruct Res.* **17**: 266.
Roth, L. E., and Shigenaka, Y. (1970). *J. Ultrastruct. Res.* **31**: 356.
Sabatini, D. D., Bensch, K., and Barrnett, R. J. (1963). *J. Cell Biol.* **17**: 19.
Schochet, S. S., Jr., Lambert, P. W., and Earle, K. M. (1968). *J. Neuropathol. Exp. Neurol.* **27**: 645.
Seidlova-Masinova, V., Malinsky, J., and Santavy, F. (1957). *J. Nat. Cancer Inst.* **18**: 359.
Shapiro, A., Vinuela, E., and Maizel, J. (1967). *Biochem. Biophys. Res. Commun.* **28**: 815.
Shelanski, M. L., and Taylor, E. W. (1967). *J. Cell Biol.* **34**: 549.
Shelanski, M. L., and Taylor, E. W. (1968). *J. Cell Biol.* **38**: 304.
Soifer, D. (1972a). *Biochim. Biophys. Acta* **271**: 182.
Soifer, D. (1972b). *J. Cell Biol.* **55**: 245a.
Stephens, R. E. (1968). *J. Mol. Biol.* **32**: 277.
Stephens, R. E. (1970a). *Science* **168**: 845.
Stephens, R. E. (1970b). *J. Mol. Biol.* **47**: 353.
Stephens, R. E. (1971). *In* "Biological Macromolecules" (S. N. Timasheff and G. D. Fasman, eds.), Vol. 4, p. 355. Dekker, New York.
Stephens, R. E., Renaud, F. L., and Gibbons, I. R. (1967). *Science* **156**: 1606.
Subirana, J. A. (1968). *J. Theor. Biol.* **20**: 117.
Sullivan, M. P., chm. (1968). *Cancer Chemother. Rep.* **52**: 453.
Tanford, C. (1961). "Physical Chemistry of Macromolecules." Wiley, New York.
Taylor, A., Mamelak, M., Reaven, E., and Maffly, R. (1973). *Science* **181**: 347.
Taylor, E. W. (1965). *J. Cell Biol.* **25**: 145.
Tilney, L. G. (1971). *In* "Origin and Continuity of Cell Organelles" (J. Reinert and H. Ursprung, eds.), p. 222. Springer-Verlag, Berlin and New York.
Tilney, L. G., and Gibbons, J. R. (1968). *Protoplasma* **65**: 167.
Tilney, L. G., and Porter, K. (1967). *J. Cell Biol.* **34**: 327.
Tu, A. T., and Friederich, C. G. (1968). *Biochemistry* **7**: 4367.
Ventilla, M., Cantor, C. R., and Shelanski, M. L. (1972). *Biochemistry* **11**: 1554.
von Wartburg, A., Angliker, E., and Renz, J. (1957). *Helv. Chim. Acta* **40**: 1331.
Wade, J., and Satir, P. (1968). *Exp. Cell Res.* **50**: 81.
Wagner, E. K., and Roizman, B. (1968). *Science* **162**: 569.
Warner, F. D., and Meza, I. (1972). *J. Cell Biol.* **55**: 273a.
Weber, K., and Osborn, M. (1969). *J. Biol. Chem.* **244**: 4406.
Weisenberg, R. C. (1972). *Science* **177**: 1104.
Weisenberg, R. C., and Timasheff, S. N. (1970). *Biochemistry* **9**: 4110.
Weisenberg, R. C., Borisy, G. G., and Taylor, E. W. (1968). *Biochemistry* **7**: 4466.
Williams, J. A., and Wolff, J. (1970). *Proc. Nat. Acad. Sci. U.S.* **67**: 1901.
Wilson, L. (1970). *Biochemistry* **9**: 4999.
Wilson, L., and Friedkin, M. (1966). *Biochemistry* **5**: 2463.

Wilson, L., and Friedkin, M. (1967). *Biochemistry* **6**: 3126.
Wilson, L., and Meza, I. (1972). *J. Cell Biol.* **55**: 285a.
Wilson, L., Bryan, J., Ruby, A., and Mazia, D. (1970). *Proc. Nat. Acad. Sci. U.S.* **66**: 807.
Wisniewski, H., Shelanski, M. L., and Terry, R. D. (1968). *J. Cell Biol.* **38**: 224.
Witman, G. B., Carlson, K., Berliner, J., and Rosenbaum, J. L. (1972a). *J. Cell Biol.* **54**: 507.
Witman, G. B., Carlson, K., and Rosenbaum, J. L. (1972b). *J. Cell Biol.* **54**: 540.
Yanagisawa, T., Hasegawa, S., and Mohri, H. (1968). *Exp. Cell Res.* **52**: 86.

ARCHITECTURE OF MAMMALIAN SPERM: ANALYSIS BY QUANTITATIVE ELECTRON MICROSCOPY

Ben Lung

DEPARTMENT OF ANATOMY
STANFORD SCHOOL OF MEDICINE
STANFORD, CALIFORNIA

I. Introduction

Historically, mammalian sperm have been objects of great fascination. Their initial discovery is usually credited to Johan Ham, a medical student, who saw motile sperm in human semen and reported his observations to the renowned Dutch microscopist Anton van Leeuwenhoek (1678). Additional observations were performed by Leeuwenhoek, who described these new organisms as small animals which possessed a distinct head and tail. Evidence that sperm are actually cells, consisting of a nucleus and cytoplasm, was first reported by Schweigger-Seidel (1865). By the latter half of the 1800's, refined dyes and improved optics permitted remarkably thorough morphological descriptions to be made by Czermak (1879), Ballowitz (1890), Meves (1899, 1901), and Retzius (1909a,b). The limited resolution of light microscopy, however, prompted Gatenby and Beams (1935) to complain that their most meticulous preparative techniques could not increase the size of sperm to the magnification at which they wished to examine them. The construction of the first high resolution electron microscope, three years later, by von Borries and Ruska (1938, 1939) has provided just the instrument that Gatenby and Beams desired.

During the years since the introduction of electron microscopy, the fine structure of mammalian sperm has been extensively studied. Many new discoveries have been reported, particularly in regard to the fine structural organization of sperm organelles and their functional relationship. The purpose of this chapter is to correlate these recent findings; in particular, emphasis will be placed upon the unique organization of the head. In addition, my own recent observations on the packing of nuclear chromatin will be presented. Earlier studies of the architecture of mammalian sperm have been summarized by Anberg (1957), Fawcett (1958, 1965), and Bishop and Walton (1960).

II. Development

Differentiation of a mammalian sperm involves many fascinating processes, among which are nuclear reorganization, development of a motile apparatus, and elimination of transitory and superfluous cellular parts. The result is an effective package capable of protecting and delivering the male genetic material for the initiation of fertilization. Parenthetically, success in the formation of a mature, fertile sperm involves the culmination of many intricate steps (for review, see Clermont, 1967; Courot *et al.*, 1970; Burgos *et al.*, 1970a). These steps are generally grouped into three major spermatogenic events: (1) division of the primary spermatogonia and their formation into spermatocytes; (2) meiosis of the spermatocytes and their subsequent transformation into spermatids; and (3) morphogenesis of the spermatid into a mature sperm.

The first two events have been reviewed by Comings and Okada

(1972a). The third, called *spermiogenesis,* has been the subject of intensive investigation and, apart from temporal and species variation, is remarkably similar in all mammals. Several species have been studied by electron microscopy, particularly with regard to differentiation of the organelles of the spermatid: cat (Burgos and Fawcett, 1955), mouse (Gardner, 1966; Sandoz, 1970), rabbit (Ploen, 1971), man (Fawcett and Burgos, 1956; Horstmann, 1961; DeKretser, 1969), rat (Brökelmann, 1963), and bandicoot (Sapsford *et al.*, 1967, 1969).

Leblond and Clermont (1952) introduced a system for subdividing the spermiogenic process on the basis of progressive changes in the acrosome, which they used as a guide for classification of the cellular events in spermatid differentiation. This system has the advantage of allowing staging of a single spermatid without the necessity for examining neighboring cells, which makes it especially suitable for correlating fine structure studies. Their system will be used here to provide a framework for understanding the component parts of the mature sperm (Fig. 1). Briefly, the elements of the system are divided into four phases, which are in turn further subdivided into stages or steps. For example, in rat spermiogenesis, the four phases include 19 stages:

Golgi Phase (stages 1–3). This phase is characterized by the appearance of several "proacrosomic vesicles," which arise from the saccular concavity of the Golgi apparatus; each contains a dense core or granule of glycoprotein (Susi and Clermont, 1970). Later, at stage 3, these small vesicles coalesce into a single large unit, called the *acrosomic vesicle,* which also contains a single acrosomic granule formed by a fusion process. This structure lies at a position near the nucleus which later becomes the anterior pole of the sperm head.

The acrosomic vesicle is closely associated with the Golgi complex and the rough endoplasmic reticulum (ER) throughout this phase. Recently, Susi *et al.* (1971) provided electron microscopic evidence, in rat spermatid, that the glycoprotein material passes from the rough ER, through the Golgi apparatus, and then into the developing acrosomic structure. Apparently, the synthesis of the carbohydrate fraction of the glycoprotein begins in the rough ER and is completed in the Golgi apparatus before release (Neutra and Leblond, 1966; Whur *et al.*, 1969). Located near the Golgi-endoplasmic reticulum complex is a densely staining, irregularly shaped *chromatoid body.* This structure is a transitory organelle which makes its earliest appearance in the primary spermatocyte, but is most easily identified in the Golgi and cap phase of spermiogenesis (Sud, 1961a,b). The origin and function of this structure is unknown. Comings and Okada (1972b) have suggested that the chromatoid body is derived from intranuclear, probably nucleolar, material.

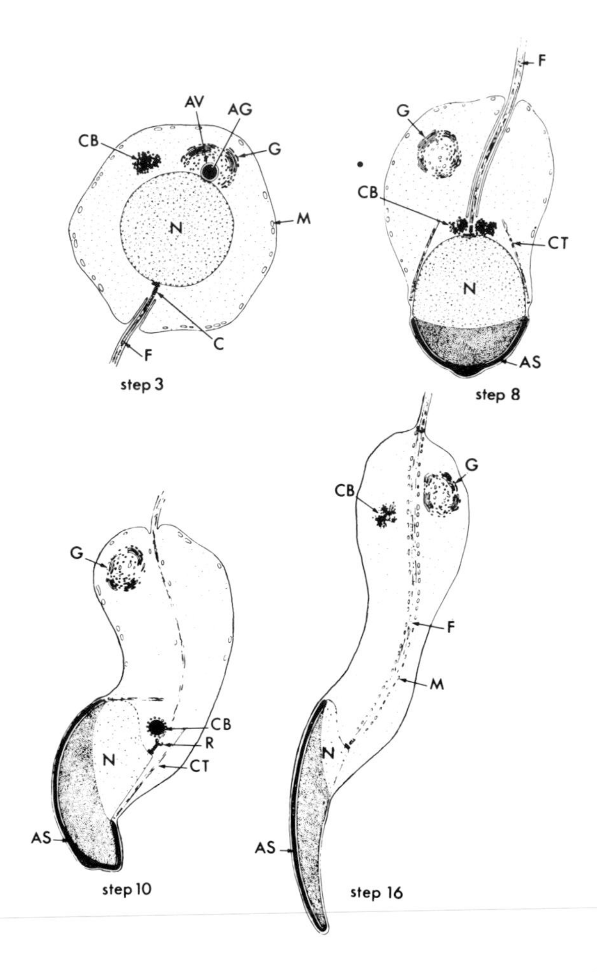

FIG. 1. A diagrammatic representation of rat spermatids in four different stages of spermiogenesis (according to the 19-stage classification of Leblond and Clermont, 1952). Illustrated are the main changes taking place in stage 3 (Golgi phase), stage 8 (early acrosome phase), stage 10 (middle acrosome phase), and stage 16 (maturation phase). AG, acrosome granule; AV, acrosome vesicle; AS, acrosomic system; C, centrioles; CT, cytoplasmic tubules (manchette); CB, chromatoid body; F, flagellum; G, Golgi apparatus; M, mitochondria; N, nucleus; R, annulus. See text for description. (From Susi and Clermont, 1970, reproduced by permission.)

On the other hand, Fawcett *et al.* (1970) maintained that it is cytoplasmic in origin. At the opposite pole of the cell, the rudiment of the tail begins to emerge from one of the centrioles and soon develops beyond the main body of the cell.

Cap Phase (stages 4–7). The Golgi apparatus becomes detached from the acrosomic vesicle and migrates backward toward the posterior part of the cell. The acrosomic vesicle during this phase enlarges and becomes flattened, forming an umbrella or cap over the anterior of the nucleus. The cytoplasm still retains a spherical outline, but at the still growing tail, a fibrous sheath starts to form.

Acrosome Phase (stages 8–14). This phase is marked by many notable changes. The acrosomic granule spreads and occupies the full compartment of the vesicle, which may now be called the *acrosome*. The nucleus begins to elongate, with a concomitant initiation of chromatin aggregation and condensation. The cytoplasm is displaced to the posterior of the cell, while internally, a system of transitory microtubules, the *manchette*, appears. This extends posteriorly into the cytoplasm and is attached anteriorly to the midnuclear region of the nuclear envelope by an annular collar called the nuclear ring (Figs. 2 and 3). The chromatoid

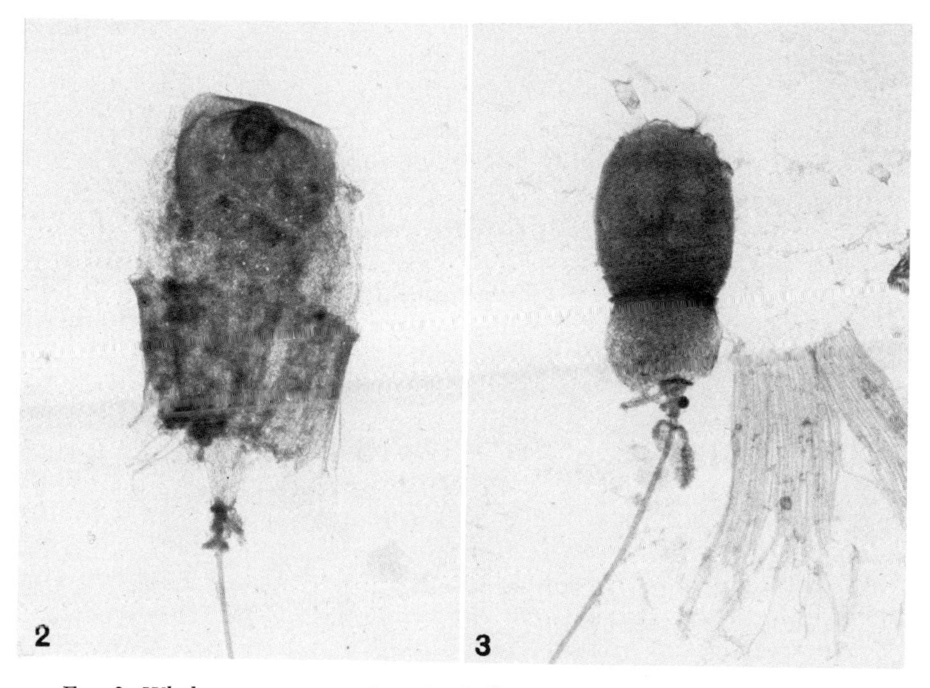

FIG. 2. Whole-mount preparation of a bull spermatid in an early stage of nuclear condensation. The acrosome granule is still prominent, and the disc of the developing acrosomal cap extends over the anterior pole of the elongated nucleus. The manchette ensheathes the caudal part of the nucleus. (Courtesy of Bahr and Gledhill, 1974.)

FIG. 3. Whole-mount preparation of a bull spermatid in a late stage of nuclear condensation. The acrosome cap covers two-thirds of the nucleus. The acrosome vesicle is not preserved. The manchette, which usually inserts at the "waist" of the head, has fallen to the side in this micrograph. (Courtesy of Bahr and Gledhill, 1974.)

body migrates to a paracentriolar position, from which it later separates and transforms into a dense sphere (Susi and Clermont, 1970; Fawcett *et al.*, 1970).

Maturation Phase (stages 15–19). This is the final and longest phase, which is keynoted by nuclear changes. The nucleus completes its aggregation and condensation, transforming the chromatin into a homogeneous structure (see Section III,F). The cytoplasm continues to be displaced. Numerous RNA staining granules (Von Ebner's) begin to arise caudally, flowing together into a single spherical aggregate called the *sphere chromatophile* (Sud, 1961b). The cytoplasm then begins to bend its caudal lobes to the anterior; these are gradually pinched off as *residual bodies*, and become embedded within the cytoplasm of the Sertoli cell, where they are phagocytosed (Dietert, 1966; Nicander, 1967; Fawcett and Phillips, 1969a). A residuum of sperm cytoplasm, called the cytoplasmic droplet, remains attached along the midpiece until ejaculation, when it is sloughed off. The final step of maturation is the release of the sperm from the seminiferous epithelium (Sertoli cell). Although little is known about this mechanism in mammals, the separation apparently takes place through an active process on the part of the Sertoli cells (Fawcett and Phillips, 1969a). Burgos *et al.* (1970b) observed, after injecting hamster testes with luteinizing hormone, the swelling of Sertoli cell cytoplasm and unfolding of the apical invaginations with a subsequent release of sperm into the tubular lumen.

III. Head

A. Cell Membrane

The cell or plasma membrane envelops the entire sperm head and continues posteriorly to cover the entire tail (Figs. 4–6). Thin sections of bull sperm reveal that the cell membrane is ultrastructurally composed of two electron dense layers separated by an intermediate light layer, a typical unit membrane which has a total width of approximately 75–80 Å (Blom and Birch-Andersen, 1965; Wooding and O'Donnell, 1971). For the most part, the sperm cell membrane resembles the membranes of other cells, but interestingly it also is highly immunological, containing a composite of surface substances (i.e., absorbed antigens), which apparently show specificity according to the genotype of the individual animal (Metz, 1967).

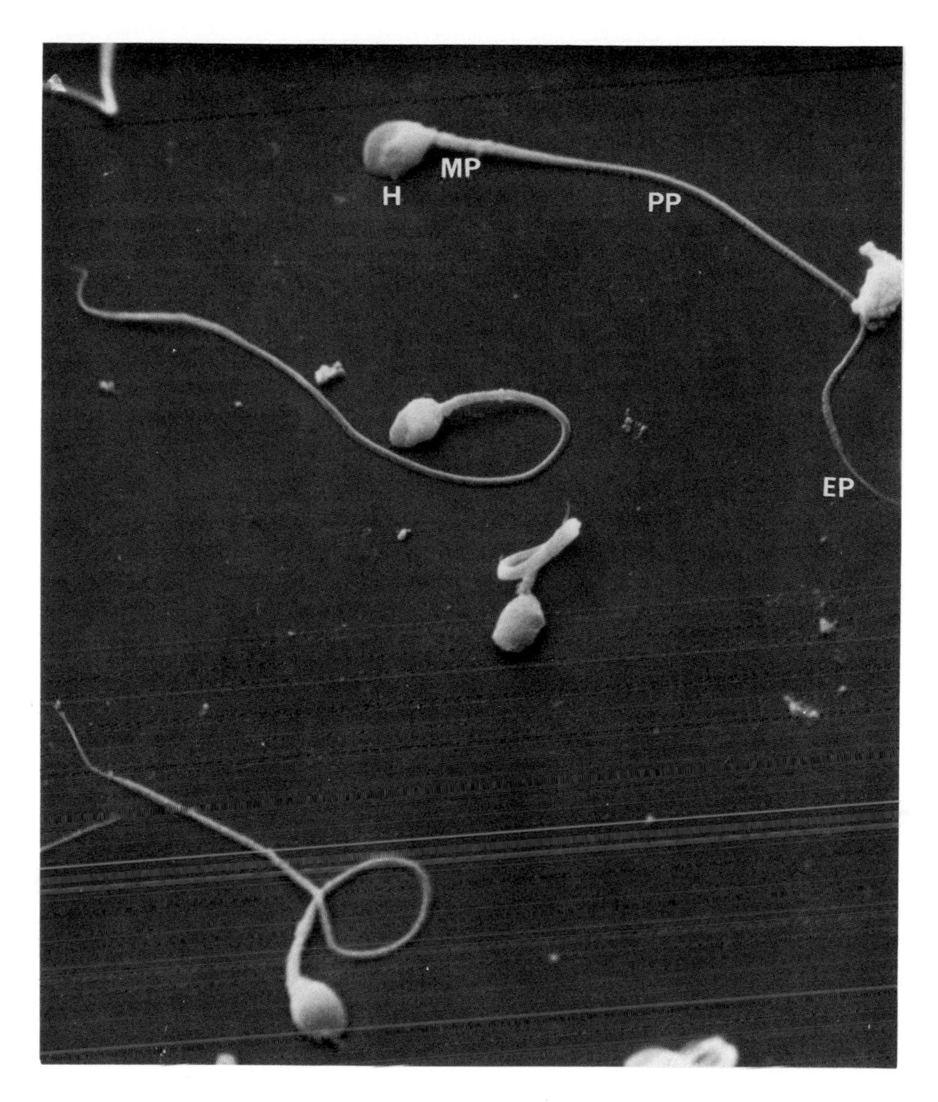

FIG. 4. Human sperm, as seen by scanning electron microscopy. The sperm head (H) is connected to the middle piece (MP), which contains the mitochondrial helix. The principal piece (PP), the longest part of the tail, is terminated by a short end piece (EP). (From Lung and Bahr, 1972.)

One notable feature, seemingly different from the membranes of other cells, is that the sperm cell membrane appears to be a fairly vulnerable structure, particularly in the periacrosomal region of the head (Fig. 5). In this region, it is consistently swollen, undulated or vesiculated,

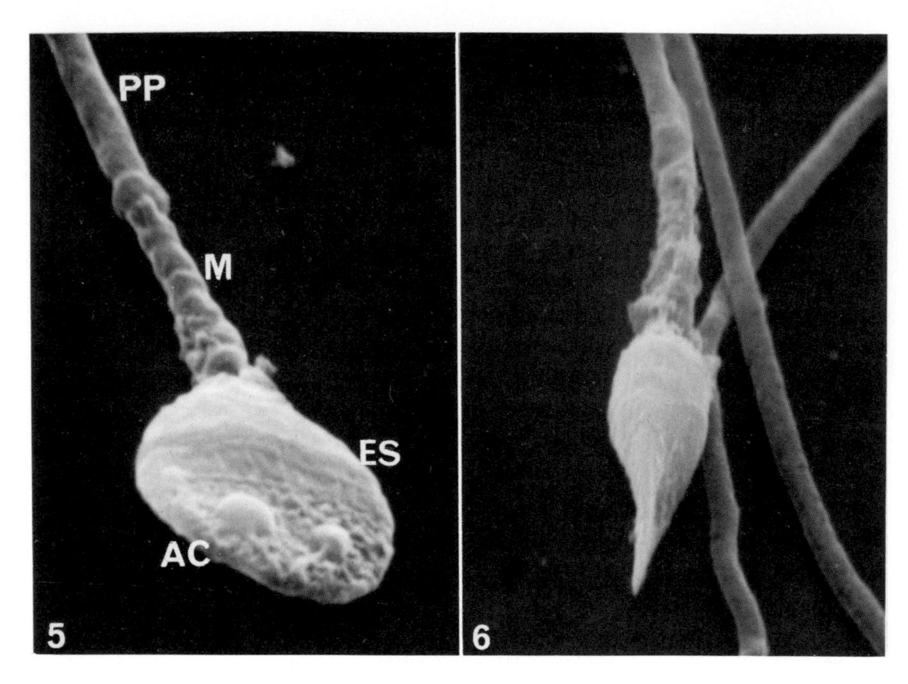

FIG. 5. Flattened view of human sperm as seen by scanning electron microscopy. The acrosome cap (AC) exhibits droplet-like elevations and is bordered posteriorly by the equatorial segment (ES), which marks the posterior border of the acrosome by two indentation grooves. The mitochondrial helix (M) of the middle piece and the principal piece (PP) are easily distinguishable. (From Lung and Bahr, 1972.)

FIG. 6. Profile view of human sperm, as seen by scanning electron microscopy, showing the raindrop-shaped head. (From Lung and Bahr, 1972).

forming a loose shroud which is detached from the underlying acrosome (Hadek, 1963; Blom and Birch-Andersen, 1965; Bedford, 1965; Nicander and Bane, 1966; Fawcett, 1970). Toward the posterior portion of the head, by contrast, the cell membrane remains closely adherent to the postnuclear cap (Saacke and Almquist, 1964a; Zamboni *et al.*, 1971).

The earliest indication of this detachment process occurs when the sperm is in the corpus and cauda epididymis, where during maturation, the cell membrane is separated free over almost the entire length of the acrosome, but does not detach from the postnuclear cap (Bedford, 1965). The obvious assumption that the detachment process is an artifact of fixation caused by osmolarity changes does not seem to be valid. Fixatives of different osmolarity do not essentially alter the loosening phenomenon in the cell membranes of epididymal sperm (Fawcett and Phillips, 1969a). For this reason, and because it is such a consistent feature in the mature sperm, several investigators (Bedford, 1965; Fawcett and Phillips, 1969a; Fawcett, 1970) have suggested that the vesiculation and detachment around the periacrosomal region should be accepted as a normal specialization of the sperm cell membrane.

In contrast, Stefanini *et al.* (1967) found that the periacrosomal cell membrane of rat sperm could be preserved reasonably intact and closely apposed to the acrosome; they employed picric acid-formaldehyde, a rapid penetrating fixative. Zamboni and Stefanini (1968) suggested that membrane vesiculation probably is not associated with sperm maturation, but may be an expression of inadequate cell preservation caused by the relatively slow penetration of fixatives, such as osmium or glutaraldehyde. This may be the reason for the commonly observed ballooning out of the cell membrane.

B. Acrosome

1. *Structure*

The acrosome arises as a secretory product of the Golgi. In the mature sperm, it consists of a saclike structure which covers the anterior one-half or two-thirds of the nucleus (Figs. 5–8). This structure, which underlies the cell membrane, is a characteristic organelle, the size and disposition

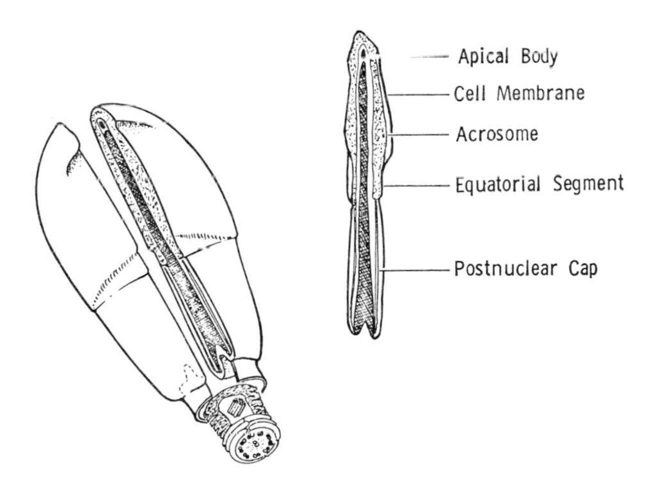

Fig. 7. Diagram summarizing the fine structural features of a bull sperm head. The connecting piece of the neck has been cut away to expose the proximal centriole. (Redrawn from Saacke and Almquist, 1964a.)

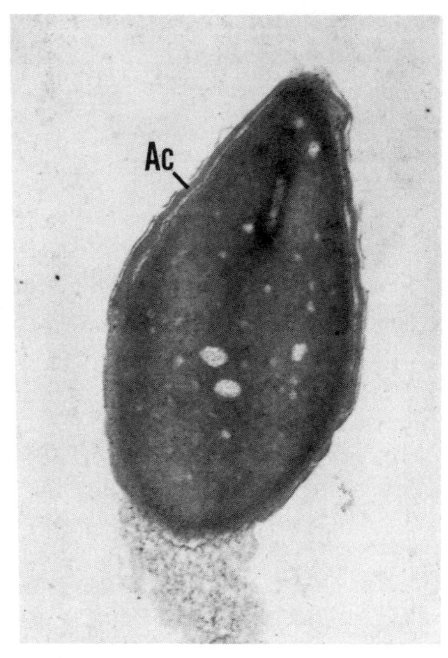

FIG. 8. Sagittal section through a human sperm head. The acrosome (Ac) surrounds the anterior two-thirds of the nucleus. In sections, the nucleus appears homogeneously opaque, interrupted by vacuoles of different sizes. ×21,200.

of which vary from species to species. Myomorph and histricomorph rodents have a particularly large acrosome that occupies a major portion of the anterior of the sperm head and extends well beyond the nucleus, forming a prominent asymmetrical segment. In other species, such as bull, rabbit, and man, the acrosome is not so well developed, but has a small apical segment that is usually symmetrical.

Thin sections reveal that the acrosome, which stains less densely than the nucleus, is rather amorphous in composition and is completely encompassed, both on the inner and outer aspects, by a continuous trilaminar unit membrane. Except at the anterior edge of the nucleus, where a pyramidal-shaped subacrosomal space is commonly observed, the unit membrane on the inner surface is closely adherent with the outer leaflet of the nuclear envelope throughout its entire length. In man, the acrosome maintains a relatively uniform thickness of about 700 Å along the length of its lateral walls, but thins to about 350 Å at its posterior edge (Pedersen, 1969). This thinning of the acrosome was first noticed by Nicander and Bane (1962a), who observed that, in boar sperm, the acrosome wall measures 600–700 Å, but is reduced to half this width to form a thin, narrow collar. They interpreted this as analogous to the equatorial segment originally observed by light microscopy. Saacke and Almquist (1964a) believed that the equatorial segment in bull sperm was probably an artifact, caused by degenerative swelling of the acrosome except in its posterior, equatorial region, and that it was absent in normal sperm. However, many subsequent investigations have demonstrated that the equatorial segment is a consistent feature in most mammalian sperm, varying only in size. This structure is now generally accepted as a specialized segment, not an artifact, of the acrosome (Bedford, 1965; Blom and Birch-Andersen, 1965; Pedersen, 1969; Yanagimachi and Noda, 1970a; Fawcett, 1970).

2. *Chemistry and Function*

Understanding of the chemical and functional aspects of the acrosome has advanced remarkably within the past few years. This is particularly true in regard to the role of the acrosome in fertilization. Recently, Piko (1969) and Bedford (1970) surveyed available chemical, functional, and structural information about the acrosome during fertilization, and the reader is referred to their reviews.

The acrosome has long been known to stain cytochemically in a way that differs from the rest of the head. Leblond and Clermont (1952) observed that periodic acid–Schiff (PAS) reacts selectively, staining the material of the acrosomal granule in developing spermatids; this property persists, the acrosome remaining stainable in the mature sperm. A PAS-positive reaction suggested to these workers that the substance responsible is a mucopolysaccharide or mucoprotein. By contrast, Onuma and Nishikawa (1963) observed that the PAS-reactive substance is strongly susceptible to lipid extraction from the acrosomes of boar, bull, and stallion, and they concluded it is more glycolipid in nature.

This discrepancy was not resolved until Hartree and Srivastava (1965) and Srivastava *et al.* (1965) developed a procedure for detaching the acrosomes from bull, ram, and rabbit sperm, by means of ionic detergent separation with Hyamine 2389. This permitted bulk isolation and a direct biochemical determination of acrosomal constituents. The isolated acrosomes were extracted with ethanol, which yielded a lipoglycoprotein complex that was positive to PAS. Even more significantly, the extract contained hyaluronidase and protease activity. In addition, when the acrosomal extracts were placed in contact with freshly ovulated rabbit eggs, the zona pellucida investing the ova were digested and dispersed from the egg surface. Boiling the lipoglycoprotein complex destroyed its dispersing capacity, which led these investigators to conclude that penetration through the zona layer probably involves a proteolytic enzyme, and that the PAS material comprising the lipoglycoprotein complex also serves as a reservoir for the acrosomal enzymes.

Austin (1948) demonstrated very early that hyaluronidase in the acrosome is required by rat sperm in order to effect passage through the cumulus oophorus. Furthermore, since hyaluronidase digestion alone does not permit the sperm head to penetrate beyond the zona pellucida, Austin and Bishop (1958) predicted that at least one other enzyme (a "zona lysin") would be necessary to dissolve that layer. However, it was not until Stambaugh and Buckley (1968, 1969) combined the Hartree and Srivastava detergent method with sucrose density gradient centrifugation that sufficient amounts of the zona lysin were purified and characterized.

The lysin, presumably similar to the one isolated by Srivastava and his co-workers, readily forms a complex with hyaluronidase, but its specific activity is 30 times greater (Stambaugh and Buckley, 1970). It also exhibits distinct proteolytic activity that is remarkably like that of trypsin, having a similar pH optimum and sensitivity to substrates and inhibitors. In the intact sperm, the lysin appears to be localized solely in the acrosome, as shown by digestion of gelatin membranes containing India ink as a marker (Gaddum and Blandau, 1970). More recently, Polakoski *et al.* (1972) were able to obtain a 250-fold purification of the trypsinlike lysin by using Sephadex chromatography. The enzyme has an estimated molecular weight of 55,000, probably in the form of a dimer which is separable into monomeric units by sodium dodecyl sulfate and mercaptoethanol. The Enzyme Commission of the International Union of Biochemistry has assigned the official name of **acrosomal proteinase** to this trypsinlike lysin. Although it behaves like trypsin in many respects, it differs by being: (1) less stable; (2) less effectively inhibited by certain trypsin inhibitors; (3) more significantly activated by calcium ions; and (4) greater in molecular weight (Zaneveld *et al.*, 1972).

A variety of other acrosomal enzymes have been demonstrated in mammalian sperm. Hartree and Srivastava (1965) presented indirect evidence for the presence of neuraminidase, as indicated by the release of sialic acid; this was later confirmed in a variety of species (Srivastava *et al.*, 1970). Yanagimachi and Teichman (1972) observed proteolytic uncoupling of silver proteinate impregnated into the acrosomes of several species; this occurred at a pH optimum lower than that of the trypsinlike enzyme. Seiguer and Castro (1972) reported an electron microscopic demonstration of aryl sulfatase activity in rat. Gordon and Barrnett

(1967) also used histochemical electron microscopy to observe acid phosphatase activity localized at the outer acrosomal membrane. Finally, Allison and Hartree (1970) detected in ram sperm acrosomal preparations the enzymes acid phosphatase, aryl sulfatase, acetylglucosaminidase, phospholipase A, and various proteases.

Although the precise significance of most of these enzymes in relation to the acrosome is not known, they collectively display a remarkable resemblance to lysosomal enzymes. Moreover, Allison and Hartree (1970) observed in a variety of species that the acrosome shows a red fluorescence, which appears to reflect a dye-binding affinity similar to that of lysosomes. On the basis of three lines of evidence—(1) ontogenetic development from the Golgi; (2) similarity in fluorescence; and (3) isolation of enzymes characteristic of lysosomes—Allison (1967) and Allison and Hartree (1970) formulated the concept that the sperm acrosome of mammals is a specialized form of lysosome containing enzymes that have evolved to facilitate fertilization.

So far, the functional role of the acrosomal enzymes has been investigated in detail only for hyaluronidase and the trypsinlike proteinase. By a process of sequential release, these have been correlated with the phenomenon of sperm-egg penetration during fertilization (Piko, 1969; Bedford, 1968, 1970, 1972; Yanagimachi and Noda, 1970b). The current view of sperm penetration into the mammalian ovum postulates an "acrosome reaction" which is divided into two parts: (1) breakdown and vesiculation of the sperm cell membrane and outer acrosomal membrane; (2) lytic digestion of the surrounding follicular cells of the egg, leading to entry of the sperm nucleus.

Barros *et al.* (1967) observed in thin sections of penetrating hamster and rabbit sperm that, as the head embeds within the outer layer of follicular cells, the sperm cell membrane and outer acrosomal membrane deteriorate and then fuse together to form a shroud of vesicles which surround the anterior part of the head. The interstices between the newly formed vesicles are believed to provide channels through which the contents of the acrosome may be released; this is presumably associated in part with early leakage of hyaluronidase to facilitate the sperm's passage through the matrix of the corona radiata (Bedford, 1968, 1970). As the sperm approaches the surface of the zona pellucida, the reacted vesicles are sloughed from the head and dissipate along with most of the acrosomal contents outside of the zona layer. In rabbit and rat, the vesiculation reaction involves the acrosomal membrane only anterior to the equatorial segment, even when the sperm head approaches the zona pellucida (Piko, 1969; Bedford, 1968). However, in hamster sperm, a greater degree of vesiculation evidently occurs, involving most of the outer membrane, including that which covers the equatorial segment (Yanagimachi and Noda, 1970b).

Dissolution of the zona pellucida is probably achieved through digestion involving the trypsinlike proteinase (Hartree and Srivastava, 1965; Srivastava *et al.*, 1965; Stambaugh and Buckley, 1968). Bedford (1968, 1970) suggested that hyaluronidase and the zona lysin may be released in sequence, or at least in two separate stages, with dissolution of the zona occurring only well after hyaluronidase digestion of the cumulus oophorus is completed. In Bedford's view, tandem removal of the two investing layers is temporally and structurally correlated with the relative disposition of the two enzymes within the acrosome. Thus, hyaluronidase would be structurally associated with the vesiculation stage and would be released first, while the trypsinlike proteinase would not be released until the head approaches the zona layer; hypothetically, the latter is bound to the inner acrosomal membrane. On the other hand, Yanagimachi and Noda (1970b) concluded from their studies on hamster sperm that the two enzymes, instead of being disposed in concentric layers, as in Bedford's hypothesis, are situated in strata layers. Hyaluronidase would occupy the anterior portion of the acrosome, lining the inner membrane as well, while the zona lysin would be situated in the posterior portion within the equatorial segment.

In hamsters, the acrosome reaction requires less than 1 hour for completion *in vitro* (Yanagimachi and Noda, 1970b). The sperm head reaches and penetrates into the zona pellucida in about 30 minutes. Next, the sperm advances through the zona in 3–4 minutes, whereupon the head comes into instantaneous contact with the surface of the egg cytoplasm (1–2 seconds). Until the entry of the head into the egg cytoplasm, penetration is mechanically aided by the vigorous beating of the sperm tail. Passage of the sperm through the zona marks the conclusion of the acrosome reaction, at which point the tail ceases to beat and only scant remnants of the acrosome remain. The final incorporation of the sperm head into the egg cytoplasm occurs by engulfment, a process described by Piko (1969) as akin to phagocytosis, i.e., the head is progressively surrounded by the spreading cell membrane and cytoplasm of the egg.

C. Apical Body

In most mammals, the subacrosomal space is occupied by an amorphous and rather enigmatic substance, which is concentrated at the anterior edge of the nucleus and completely fills the triangular cavity of the subacrosomal space throughout most of its length (Fig. 7). This substance is called the apical body or "perforatorium." (The name perforatorium is a confusing and unfortunate term, because it ambiguously implies a function and has also been used to designate the anterior apex of the acrosome in some species.)

The apical body was first described by Moricard (1961) in rabbit sperm, and it was later observed in ram, dog, cat, guinea pig, and bull sperm (Saacke and Almquist, 1964a; Blom and Birch-Andersen, 1965; Nicander and Bane, 1966). However, other investigators were unable to find any appreciable material within the subacrosomal space of guinea pig and bat; they questioned the existence of any formed structure that could be called an apical body (Fawcett, 1965; Fawcett and Ito, 1965). Factors having to do with chemical fixation during specimen preparation may be responsible for the inability of these workers to observe the structure. Subsequently Gordon (1969) showed that the presence of the apical body could be demonstrated when the sperm head was specifically treated with ethanolic phosphotungstic acid, whereas it was not revealed by treatment with osmium. Moreover, it persists as a definite structure, retaining its morphology even after removal of the acrosome and subacrosomal membranes with detergents (O'Donnell *et al.*, 1970).

The development of the apical body is fundamentally the same in most mammals, and has been studied in varying detail by several investigators (Bane and Nicander, 1963; Hopsu and Arstila, 1965; Sandoz, 1970; Bedford and Nicander, 1971; Lalli, 1971; Franklin and Fussell, 1972). This structure first arises during cap phase as a thin layer of fine flocculent material, which in certain species may be interrupted in succeeding stages by transitory whorls of membranes (Hopsu and Arstila, 1965; Sandoz, 1970; Franklin and Fussell, 1972). The flocculent material persists throughout most of spermiogenesis, while the membrane whorls, which on occasion appear continuous with the nuclear envelope, are replaced during maturation phase by a dense accumulation of coarse granular material (Franklin and Fussell, 1972). In the last few steps of spermiogenesis, both the fine flocculent and coarse granular material fuse and condense into an amorphous structure that subsequently forms the mature apical body.

The functional significance of the apical body in mammals is still obscure and has been a subject of much controversy which remains to be resolved. Because of its strategic position, Hadek (1963) suggested that the apical body may be analogous to the perforatorium rod observed in several invertebrate sperm, such as sea urchin and mussel; this structure mechanically aids sperm penetration into the ovum (Dan, 1967). Other investigators, while unable to detect evidence of mechanical perforation by the apical body, believe that it may contain enzymes that enhance sperm penetration (Bedford, 1970). Still others deny that the structure plays any significant role in fertilization and suggest that the apical body may act only as a cementing substance to bind the acrosome to the underlying nucleus (Fawcett and Phillips, 1969a; Fawcett, 1970).

D. Postnuclear Cap

Positioned immediately posterior to the equatorial segment, the postnuclear cap consists of a cytoplasmic sheath of dense material located between the cell membrane and the nuclear envelope, extending caudally to, but not continuous with, the basal plate (Fig. 6). The term "postnuclear cap" takes its origin from light microscopy, in which the structure was demonstrated by its strong affinity to silver salts (Gatenby and Wigoder, 1929). Fawcett (1965) proposed that the term should be abandoned because it is not descriptive of the actual structure, as observed in thin sections: i.e., the component material of the postnuclear cap does not form a closed, continuous cap, but is open and discontinuous at its posterior aspect. Instead, he recommended that the term "postacrosomal dense lamina" would be more appropriate (Fawcett and Phillips, 1969a). Many investigators have since adopted this suggestion, and have discarded the earlier, less adequate term. However, in this review the original designation will be retained, in order to avoid confusion with the already widespread usage in the older, prevailing literature.

The postnuclear cap is not a true membrane, although it may be mistaken for one because it is usually so closely attached to the cell membrane that it is difficult to resolve as a separate structure. Although current investigators of sperm ultrastructure are in general agreement concerning the presence of a postnuclear cap as a distinct structure, there is considerable difference of opinion about the details of its fine structure. Blom and Birch-Andersen (1965) interpreted the postnuclear cap in bull sperm as a sheath, composed of a palisade of microtubules. Koehler (1966, 1970) observed in bull and rabbit sperm prepared by freeze-etching that the postnuclear cap had, instead, surface striations. Rahlmann (1961) reported it to be porous in bull sperm. Nicander and Bane (1962a) observed in boar sperm that the structure was rather amorphous, displaying a marked affinity to heavy metals, including phosphotungstic acid as well as silver salts. In rabbit, Bedford (1964) found that the postnuclear cap had a spongy appearance when stained either with lead or with uranyl acetate followed by potassium permanganate.

Human sperm, on the other hand, have a weakly developed postnuclear cap which is relatively inconspicuous, consisting of diffuse and almost undetectable material (Bedford, 1967; Pedersen, 1969; Zamboni *et al.*, 1971).

More recently Wooding and O'Donnell (1971), in a detailed reinvestigation of bull sperm head membranes, observed that the composition of the postnuclear cap material is highly dependent upon the type of fixative employed. Fixation with potassium permanganate or glutaraldehyde followed by potassium permanganate, resulted in the postnuclear cap appearing as an amorphous layer, which was closely applied to the cell membrane but separated from the nuclear envelope. On the other hand, fixation with glutaraldehyde followed by osmium resulted in the postnuclear cap taking on an entirely different appearance. Numerous longitudinal striations were found on its surface, which were also randomly interdispersed by small pores at the upper half.

Development of the postnuclear cap begins rather late in spermatid formation, at about the time when the nucleus has already become elongated and flattened. As observed by light microscopy, the postnuclear cap forms from an argentophilic band positioned along the postacrosomal border in bull and boar; this extends from the nuclear ring backward to the caudal pole of the head (Hancock, 1957; Hancock and Trevan, 1957). Pedersen (1970) observed, in thin sections of human spermatids, that an electron dense material lies medial to the cell membrane in the region of the nuclear ring and presumptive equatorial segment; this moves backward, forming the postnuclear cap structure. Development of this structure apparently is completed by the time the sperm reaches the epididymis.

At present the functional significance of the postnuclear cap is not known. Fawcett and Ito (1965) postulated, from their studies of the epididymal sperm of bats, that it may act merely as a cementing substance for the postacrosomal cell membrane. This suggestion, however, may not be the entire explanation for the structure's function. Recent studies have demonstrated that, upon entry of the sperm into the egg ooplasm, initial contact involves adhesion and fusion between the mid-posterior surface of the sperm head and the vitellus (Barros and Franklin, 1968; Piko, 1969; Bedford, 1970). After disappearance of the cell membrane during the fusion process, the ooplasm is first exposed specifically to the material of the postnuclear cap. This has suggested to Bedford (1972) that the postnuclear cap may also play a significant role either in activation of the egg or incorporation of the sperm nucleus into the ooplasm.

E. Nuclear Envelope

The nuclear envelope of the sperm head forms the innermost cellular investment surrounding the nucleus. It consists of two unit membranes which, depending upon the fixative employed, either are divided into two separate layers or are fused to appear like two unit membranes with a common central leaflet (Wooding and O'Donnell, 1971).

In general, the nuclear envelope of sperm is an unobtrusive structure, closely bound to the perinuclear surface and very often difficult to resolve. However, at the base of the nucleus it is reflected away from the main nuclear body to form a distinct diverticulum called the *redundant nuclear envelope* (Fig. 9). This unusual feature has attracted the particular attention of many investigators and has been characterized in a variety of mammalian species, such as bat (Fawcett and Ito, 1965; Wimstatt *et al.*, 1966), boar (Nicander and Bane, 1962b), bull (Blom and Birch-Andersen, 1965; Nicander and Bane, 1966; Wooding and O'Donnell, 1971), hamster (Fawcett, 1970; Yanagimachi and Noda,

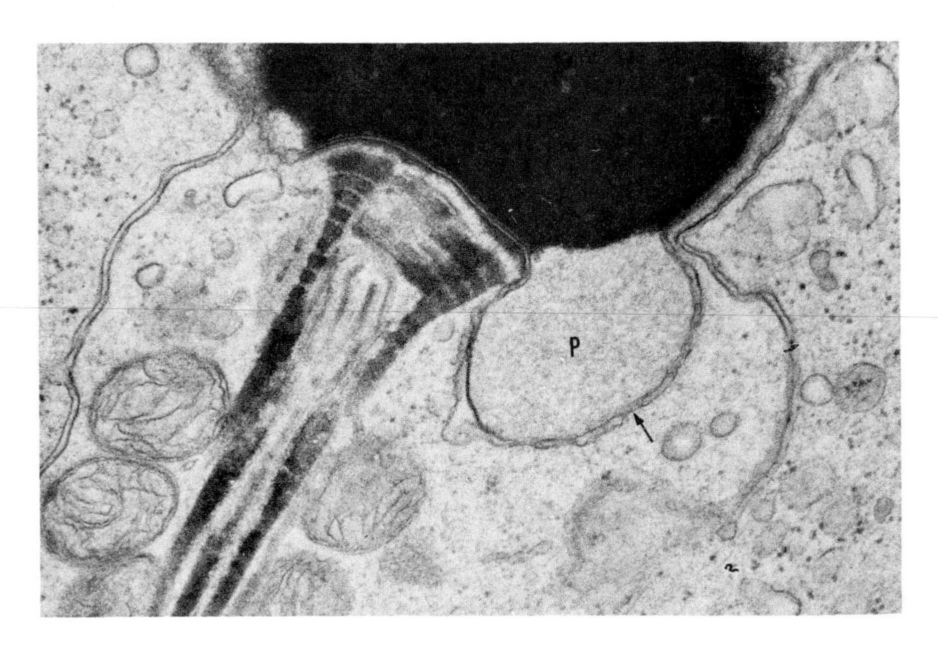

Fig. 9. Longitudinal section through the neck region of a late spermatid of rhesus monkey. The redundant nuclear envelope, containing nuclear pores (arrow), is reflected from the base of the nucleus back into the neck region of the sperm. The chromatin appears opaque, except where nuclear protrusions (P) extend posteriorly into the expanded space of the nuclear envelope. (Courtesy of Franklin, 1968. Reproduced by permission.)

1970a), guinea pig (Fawcett, 1965; Nicander and Bane, 1966), man (Bedford, 1967; Zamboni *et al.*, 1971), monkey (Bedford, 1967; Franklin, 1968; Zamboni *et al.*, 1971), stallion, sheep, dog, and rabbit (Nicander and Bane, 1966).

According to Fawcett (1965), the redundant portion of the nuclear envelope arises at the time of chromatin condensation, and its formation is correlated with a reduction of the nuclear volume during spermiogenesis; the excess envelope is not absorbed, but extends back toward the neck. Morphology of this region is quite variable among the different species. In some mammals, such as rhesus monkey, bull, and marmoset, the redundant nuclear envelope bends back upon itself to form a simple fold (Franklin, 1968; Blom and Birch-Andersen, 1965; Rattner and Brinkley, 1971). In others, such as bat, bush baby, and Russian hamster, it is a much more elaborate structure, forming an extensive scroll of membranes (Fawcett and Ito, 1965; Bedford, 1967; Fawcett, 1970). Moreover, in many mammals the redundant nuclear envelope is accompanied by nuclear extensions, which protrude into the evaginated spaces to form "basal knobs" of chromatin material (Blom and Birch-Andersen, 1965). These protrusions are less dense than the main body of the nucleus, hence they are composed of incompletely aggregated and condensed chromatin (Nicander and Bane, 1966; Franklin, 1968; Wooding and O'Donnell, 1971). Nuclear pores, or annuli, are also restricted to the redundant portion, and as a rule do not occur where the envelope is closely applied to the nucleus. These pores measure about 600 Å in diameter in monkey (Franklin, 1968) and 1000 Å in bull (Wooding and O'Donnell, 1971); they apparently have an annular structure, with a central density extending across the opening.

The functional significance of the redundant nuclear envelope is not known, but it is unlikely to be simply a vestige of nuclear condensation. Fawcett (1970) has speculated that in some mammals, where the surplus envelope has retained an extensive scroll-like structure, it may represent an emergency endogenous energy reserve in the absence of exogenous carbohydrate substrates. On the other hand, Wooding and O'Donnell (1971) postulated that it may function as a shock absorber for the violent thrashing movement of the tail. Horstmann (1961) considers that the excess nuclear membrane aids in removing unwanted nuclear material to the cytoplasm. This interesting possibility has some support from more recent studies in nonmammalian spermatids of the polychaete annelid (Potswald, 1967), earthworm (Anderson *et al.*, 1967), dragonfly (Kessel, 1966), and tidepool sculpin (Stanley, 1966, 1969), where the excess nuclear envelope often contains nuclear material and is pinched off from the surface of the nucleus.

F. Nucleus

1. *Structure*

Electron microscopy has demonstrated consistent morphological differences between the sperm nucleus and nuclei of somatic cells. Architecturally, the nucleus of the mammalian sperm head contains a much more condensed or densely packed chromatin mass, which in thin sections is homogeneously opaque, and is usually devoid of resolvable detail except for small interruptions or irregular clear areas referred to as *nuclear vacuoles* (Fig. 8). This density or opacity of the nuclear substance in mammalian sperm is well known (Fawcett, 1958, 1970; Blom and Birch-Andersen, 1965; Nicander and Bane, 1966; Yanagimachi and Noda, 1970a), and is believed to be derived from the remarkable molecular reorganization that the spermatid nucleus undergoes. A highly aggregated and condensed chromatin results, that is metabolically inert and surprisingly resistant to digestion by enzymes or disruption by physical and chemical methods (Fawcett, 1958).

However, difficulty in detecting discrete substructure in the nucleus of a mature sperm does not imply that one does not exist. Recent investigations indicate that the nuclear substructure may be too closely packed to be resolved in ordinary electron micrographs, or possibly it may be obscured by the usual methods of specimen preparation. Unlike other cells, the sperm nucleus in mammals specifically contains a high percentage of cystinyl disulfide cross-links, which in conjunction with the nucleoproteins stabilize and aggregate the chromatin into a highly condensed state (Bril-Petersen and Westenbrinik, 1963; Calvin and Bedford, 1971). By rupturing of the cross-links, the chromatin may be dispersed into fibrous units which are not normally observed in the intact nucleus (Lung, 1968, 1972). Disaggregated chromatin fibers are arranged into thick bundles, which in a completely dissociated state consist of unit fibers having a diameter of about 140–240 Å. In man, the DNA is tightly packed within each unit fiber; dry mass profiles obtained by quantitative electron microscopy indicate that there may be a coiled coil configuration (see Section III,F,3).

Freeze fracturing performed on bull and rabbit sperm has also demonstrated a high degree of nuclear order at the supermacromolecular level, where none can be observed by the usual techniques of thin sectioning (Koehler, 1966, 1970; Plattner, 1971). By this method, thin nuclear laminations have been observed forming 100 Å plates stacked parallel to the flat surface of the head and spaced 250–300 Å apart, which display a more complex organization than that found in human sperm

(Koehler, 1966, 1972). Koehler (1970) suggested that these plates are produced in a manner very similar to that found in certain invertebrate spermatids, such as grasshopper and locust, in which the chromatin fibers aggregate into lamellae or plates, which then become obscured in sections of mature sperm. Later birefringence investigations seem to reinforce this suggestion (Bendet and Bearden, 1972; Bearden and Bendet, 1972). In bull sperm nuclei, there is form birefringence indicating the presence of lamellar structures, but this is lost when the nuclear disulfide linkages are ruptured. At the same time, the nuclei do not demonstrate a significant alteration in melting temperature. This implies that the splitting of -S-S- linkages separates the platelike structures into a finer component, presumably fibers, but does not effect a separation of the DNA from the nucleoproteins, which remain complexed by salt bridges.

Nuclear vacuoles usually are randomly interdispersed within the sperm nucleus of most mammals. They are irregular in outline, vary greatly in size and shape, and, unlike true cytoplasmic vacuoles, do not possess a limiting membrane. In man, the vacuoles are quite common, often attaining a relatively large diameter of 0.8 μ, particularly in the anterior portion of the nucleus. Smaller vacuoles, which are more frequent, measure 0.05–0.1 μ in diameter and are randomly located in the nucleus. In sperm of other species, such as bull, the vacuoles are usually quite small, measuring approximately 0.1 μ (Blom and Birch-Andersen, 1965). The cavities of nuclear vacuoles are usually empty, but occasionally some amorphous material may be found. In human sperm, membranous whorls also have been observed (Pedersen, 1969; Zamboni *et al.*, 1971), and in hamster spermatids, microtubular structures have been seen (Hadek, 1969). The nuclear vacuoles do not seem to have a specific functional significance. However, Fawcett (1958) has pointed out that they are of such common occurrence that they are probably not artifacts, but local defects resulting from the condensation process in the nucleus.

The structural changes that the sperm nucleus undergoes during spermiogenesis are essentially the same among the different varieties of mammal (cat: Burgos and Fawcett, 1955; man: Fawcett and Burgos, 1956; Horstmann, 1961; DeKretser, 1969; bandicoot: Sapsford *et al.*, 1969; Sapsford and Rae, 1969; mouse: Gardner, 1966; wooly opossum: Phillips, 1970; rabbit: Ploen, 1971). These changes involve a process of nuclear condensation which begins during the acrosome phase of spermiogenesis, when the nucleus becomes flattened and elongated; at this time the chromatin proceeds to aggregate and condense into coarse units. Clumping of the chromatin then occurs, which is synchronously associated with a marked reduction in the nuclear volume. By the end of the maturation phase, the chromatin is completely condensed, and in thin-sectioned material it appears to form a homogeneous, electron-dense mass in the mature sperm.

The traditional interpretation of substructure in the spermatid nucleus during condensation has been one that describes the chromatin as small granules. Indeed, in sections the chromatin elements do have a rounded, punctate profile. However, upon closer examination, these circular outlines may also be interpreted as the cut ends of tightly coiled fibers, rather than true granules (Lung, 1972).

Recent investigations, employing whole-mount electron microscopy, have now done much to clarify this question. In unsectioned male meiotic cells of man, the chromosomes are completely fibrous in nature and have a structure that is basically similar to that found in somatic cells (Comings and Okada, 1971). Bahr and Gledhill (1974) have also found recently that the chromatin elements in the spermatids of the bull are constructed completely of fibers, arranged in a network with little or no architectural order (Figs. 2 and 3). This work provides compelling evidence for the fibrous nature of the spermatid nucleus, which is maintained in the mature sperm in a highly aggregated and condensed form.

2. *Chemistry*

In contrast to the state of knowledge about deoxynucleoproteins in somatic nuclei, present understanding of the DNA-nucleoprotein complexes in the sperm of mammals is still at a relatively primitive stage. One reason for this is that isolation by the usual extraction methods, either of DNA or of nucleoprotein, has been hampered by insoluble disulfide cross-links formed within the cystinyl residues of the nucleoprotein. Because of this chemical resistance, many early investigators were able to obtain only crude extracts of the sperm nuclear components, and these also apparently included several contaminants.

Borenfreund *et al.* (1961) were able to develop a successful extraction procedure, employing a disulfide reducing agent, mercaptoethanol followed by trypsin digestion; their technique permitted isolation and recovery of an estimated 96–99% of the sperm DNA. This DNA showed a high degree of regularity in base composition and had a buoyant density of 1.70, implying that it is about the same as DNA from other mammalian tissues. Moreover, these workers found that the sperm DNA had a very high melting temperature, which also indicates a high degree of nativeness. On the average, a mammalian sperm contains about 3×10^{-12} gm of DNA, approximately half the content found in somatic cells (Bishop and Walton, 1960). However, in the sperm it is considerably more concentrated, making up about 43–48% of the deoxynucleoprotein composition (Leuchtenberger *et al.*, 1956; Dallum and Thomas, 1953).

Hendricks and Mayer (1965a,b), with a modified mercaptoethanol procedure followed by acid precipitation, found in bull and boar sperm that the nucleoprotein could be isolated from the complexing DNA. This protein was very rich in arginine, but was distinct from typical protamine or histone. Instead, it resembled a tough basic keratinoid, especially with regard to its insolubility and high content of cystine. Bril-Petersen and Westenbrinik (1963) also isolated from the arginine-rich protein of bull sperm over 6% cystine, and found that the protein bore little resemblance to the somatic type lysine-rich histones. In fact, little or no lysine was found in their extraction fractions, while the arginine content measured greater than 35%. Interestingly, the molar ratio of arginine to DNA phosphoric acid was about 0.9, while the ratio of total basic amino acids to DNA phosphoric acid was only slightly higher at about 1.0; this suggests that most of the arginine is complexed to the DNA. On the basis of differences in extraction rates, these workers also suggested that the high protein sulfur observed in their fractions was predominantly in the form of intranuclear cross-links connecting the arginine-rich proteins.

Extending the study of Bril-Petersen and Westenbrinik, Coelingh *et al.* (1969) found that the sperm nucleus consists of a network of identical arginine-rich, nucleoprotein molecules which are mutually linked by cystine disulfide bridges. The molecular weight of the molecule is about 6200, composed of 47 amino acid residues; of these, 24 are arginine, 6 cysteine, 3 threonine, and 10 other amino acids occur once or twice in the molecule. The amino acids lysine, proline, aspartic acid, and methionine are completely absent. The amino-terminal residue is alanine, and the carboxy-terminal residue is glutamine. More recently, Coelingh *et al.* (1972) have determined the complete amino acid sequence of the arginine-rich protein in bull sperm, using partial proteolytic digestion and chemical degradation. The sequence is as follows:

$$\mathrm{H_2N\text{-}Ala\text{-}Arg\text{-}Tyr\text{-}Arg\text{-}(Cys)_2\text{-}Leu\text{-}Thr\text{-}His\text{-}Ser\text{-}Gly\text{-}Ser\text{-}Arg\text{-}Cys\text{-}(Arg)_7\text{-}Cys\text{-}(Arg)_6\text{-}}$$
$$\mathrm{Phe\text{-}Gly\text{-}(Arg)_6\text{-}Val\text{-}Cys\text{-}Tyr\text{-}Thr\text{-}Val\text{-}Ile\text{-}Arg\text{-}Cys\text{-}Thr\text{-}Arg\text{-}Gln}$$

Examining the distribution of arginine and cysteine through the molecule after partial proteolytic digestion reveals that the central portion is very basic, two of three residues being arginine; by contrast, at least half of the cysteine residues are situated in the more terminal portions of the molecule. This suggests that most of the central part of the nucleoprotein molecule is apparently associated with DNA in the form of salt bridges, while the more terminal parts may be involved in forming cross-links between protein molecules (Coelingh *et al.*, 1972). The molecule contains a paucity of α helical segments when displayed in the form of helical wheel projections (Schiffer and Edmundson, 1967) which implies that the secondary structure is primarily in an extended or folded form, similar to sperm protamines (Wilkins, 1956).

Investigations into the process by which somatic histones are replaced by the arginine-rich proteins have largely been carried out by histochemical and autoradiographic methods. Surprisingly, this transition process occurs rather late in spermiogenesis, and in a relatively brief period of time. Vaughn (1966) observed in mouse that the lysine-rich, somatic histones are replaced rapidly and completely by arginine-rich proteins within a period which lasts about 66 hours and begins in the late acrosome phase (stage 13), when the nucleus is in a stage of elongation. Synthesis of the new arginine-rich proteins is initiated only slightly before this protein transition; according to Monesi (1964, 1965), this synthesis, which begins abruptly at stage 11 of the acrosome phase, continues vigorously in conjunction with the protein transition until stage 15–16 of the maturation phase, when it suddenly stops as quickly as it began. Meanwhile, the lysine-rich histones are continuously displaced to the spermatid cytoplasm, where they are localized in the sphere chromatophile, a large RNA-containing mass which is later cast off as a component of the residual bodies (Vaughn, 1966).

The transition in sperm nucleoprotein coincides with the beginning of fibrillar aggregation and nuclear condensation; of particular interest, this process is also associated with progressive stabilization of the deoxynucleoprotein. Gledhill *et al.* (1966) observed that, as the protein transition and chromatin condensation take place, there is a reduction in DNA-Feulgen stainability in the absence of a reduction of DNA. This lowered capacity of the deoxynucleoprotein to bind dye apparently is coupled to an increase in protein-bound arginine groups, which increases the total basicity of the nucleoproteins and results in a decrease of the available DNA phosphate groups. This finding indicates that there is an increase in the strength of electrostatic binding between spermatid DNA and its nucleoproteins; with the onset of nuclear condensation, this change is closely correlated both with a progressive stabilization of the DNA against heat denaturation (Ringertz *et al.*, 1970; Gledhill, 1971) and with a marked reduction in the ability of DNA to bind ^{3}H-labeled actinomycin D (Darzynkiewicz *et al.*, 1969).

Formation of disulfide linkages apparently also contributes to the stability of the deoxynucleoprotein. Loir (1970) observed that cystine-^{35}S increases simultaneously with arginine-^{3}H as they are incorporated into the sperm nucleus, suggesting that the cystine may contribute to the stabilization of the chromatin. Moreover, Calvin and Bedford (1971) observed that an increase in formation of cystinyl —S—S— bonds in the spermatid is accompanied by an increase in resistance of the nuclear chromatin to lysis.

After maturation, sperm nucleoprotein evidently does not undergo further modification until after fertilization. By autoradiographic labeling, Kopečny (1970) showed that arginine-^{14}C is preserved within the nucleus of the mature mouse sperm, but cannot be detected in the male pronucleus after fertilization. Disappearance of the label implies that the arginine-rich protein is no longer conserved beyond male pronuclear formation. This situation may be similar to that found in other animals (snail, Bloch and Hew, 1960; *Drosophila,* Das *et al.,* 1964), in which the sperm basic proteins undergo a second transition after fertilization, namely, loss of arginine-rich proteins and acquisition of another type of nucleoprotein. In mammals, unfortunately, the nature of this change to a new nucleoprotein is obscure; whether it involves direct replacement with a somatic type of histone, or temporary replacement with a transitory protein, remains to be resolved.

The RNA content of mammalian sperm was investigated by several early workers, who detected only varying trace amounts (Vendrely and Vendrely, 1948; Leuchtenberger *et al.,* 1952; White *et al.,* 1953). More recent studies indicate that the nucleus of a mature sperm lacks virtually any detectable RNA; it is present in detectable amounts only during sperm development (Monesi, 1964, 1965; Utakoji, 1966). As shown by autoradiographic incorporation of uridine-^{3}H, intense synthesis of RNA occurs during the meiotic prophase (Monesi, 1965); this RNA is DNA-like in composition and is turned over rapidly (Muramatsu *et al.,* 1968). Unlike the corresponding oocytes, spermatocytes lack nucleolar incorporation of uridine-^{3}H, suggesting that although ribosomal RNA from an earlier synthesis may be present, it is not synthesized detectably during meiosis (Utakoji, 1966; Muramatsu *et al.,* 1968). Synthesis of DNA-like RNA in the spermatid nucleus completely ceases in early spermiogenesis, soon after the second meiotic division, and no further incorporation of uridine-^{3}H can be detected after the acrosome phase (stage 8–9, Monesi, 1965). This arrest of RNA synthesis at the beginning of nuclear elongation coincides with a cessation of nuclear protein synthesis, except for the ensuing synthesis of arginine-rich proteins (Monesi, 1971). Interestingly, this synthesis appears to be supported by a long-lived messengerlike RNA, i.e., actinomycin D, when administered to spermatocytes at a dose that inhibits RNA synthesis, does not prevent protein synthesis in the spermatids (Monesi, 1967). Elimination of residual RNA from the nucleus is accomplished mainly by breakdown and transfer to the spermatid cytoplasm. By late spermiogenesis, the accumulated cytoplasmic RNA is concentrated in the residual bodies, and is eventually eliminated from the cell (Monesi, 1965).

3. *Chromosome Packing*

a. Chromatin Disaggregation and Decondensation. A variety of sulfhydryl-containing reducing agents have the capacity to disaggregate and decondense nuclear chromatin in mammalian sperm, by disulfide reduction or disulfide exchange (Table I). This rupturing of disulfide bonds in proteins is highly specific for thiols, such as thioglycolate (Boyer, 1959). However, alkalinity is also necessary for rapid nuclear disaggregation and decondensation. At pH 9.0, the SH groups of thioglycolate are only partially ionized, and the reduction reaction requires several minutes to decondense the sperm nucleus. Increasing the pH (or concentration) of the solution causes almost instantaneous disruption of sperm heads. Recently Calvin and Bedford (1971) reported that the

TABLE I
HUMAN SPERM NUCLEAR EXPOSURE WITH DIFFERENT SULFHYDRYL-CONTAINING REDUCING AGENTS[a]

Reducing agent	pH 7	pH 8	pH 9	pH 10	pH 11
Thioglycolate					
0.5 *M*	+	+++	+++	++++	++++
0.25 *M*	−	+	++	+++	+++
Mercaptoethanol					
0.5 *M*	−	+½	++	+++	+++
0.25 *M*	−	−	+	+	++
Mercaptopropionate					
0.5 *M*	−	½	+	+++	++++
0.25 *M*	−	−	½	+	+++
Cleland's reagent (dithiothreitol)					
0.5 *M*	−	−	+	+++	++++
0.25 *M*	−	−	½	++	+++

[a] Approximate percentage of nuclear unpacking in human sperm after treatment with different sulfhydryl-containing reducing agents (as determined by visual observations with phase-contrast microscopy). Molar concentrations represent final concentrations of the solutions. Length of observation time was 3 minutes. −, no visible change; +, nuclear exposure of 10–20% sperm; ++, nuclear exposure of 50% of sperm; +++, nuclear exposure of 80% of sperm; and ++++, complete nuclear exposure of all sperm.

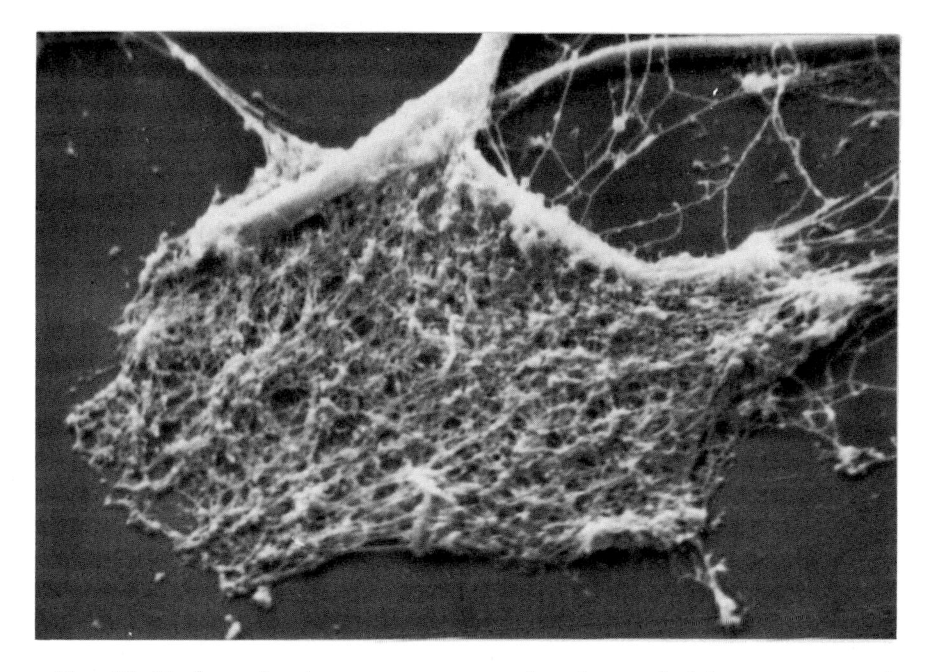

FIG. 10. Nucleus of a human sperm exposed and unpacked by treatment with alkaline thioglycolate, as seen by scanning electron microscopy. The nucleus is composed of chromatin made up of fibers, which are arranged in a network after thioglycolate treatment. (From Lung and Bahr, 1972.)

sperm chromatin of various mammals can also be decondensed with Cleland's reagent dithiothreitol. They found that the rupturing of disulfide bonds is most controllable at pH 9.0, when the Cleland's reagent is half ionized (Zahler and Cleland, 1968); maximal dissolution of the sperm head occurs at pH 10.5, when the reducing agent is approximately 95% ionized.

After specific rupturing of disulfide bonds, the chromatin of human sperm is observed to be strikingly fibrillar, arranged in an irregular network without any indication of coherent order (Fig. 10). The fibers unpack progressively during the disulfide reduction, and the nucleus undergoes a remarkable change from a condensed to a diffuse state. Individual fibers in the diffuse state frequently possess thickenings or "nodes" approximately twice the fiber diameter and distributed more or less randomly along the fiber axis (Lung, 1968). Although the functional significance of the nodes is not known, they appear ultrastructurally to be part of the chromatin fiber, rather than contaminants attached during the preparative procedures. In order to confirm this point, the relative dry mass distribution within the nodes and the internodal

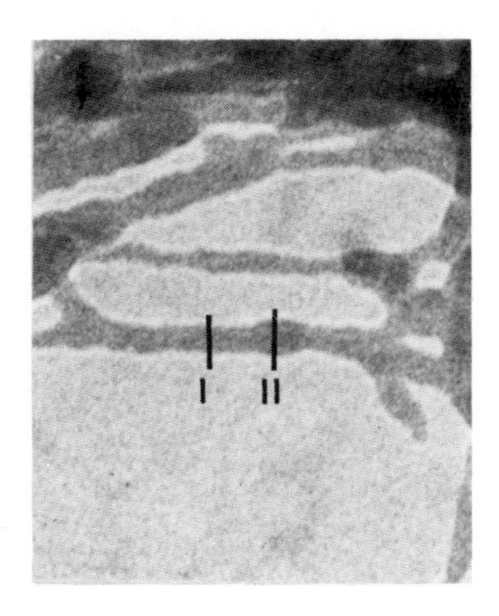

FIG. 11a. High magnification micrograph of chromatin fiber from human sperm, showing a nodelike thickening. High resolution densitometric scans were performed on the fiber cross section (line I), and on the nodelike thickening (line II). ×94,700.

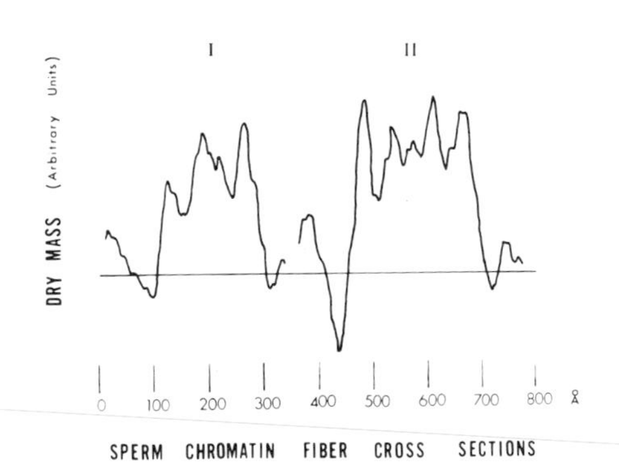

FIG. 11b. High resolution densitometric tracings of the fiber cross sections shown in Fig. 11a. Areas beneath the curves represent the cross-sectional dry mass of the fiber (curve I = 487 units), which has a density approximating the density determined from the dry mass of the nodelike thickening (curve II = 840 units).

fibers has been determined by cross-sectional scans with a sensitive, high resolution microdensitometer (Fig. 11). After the relative dry mass is obtained from the scans, the relative density in arbitrary units can be estimated on the basis of geometric calculations for a cylinder. The nodal and fiber density values calculated in this way are in good approximation to each other. This result tends to eliminate such macromolecular contaminants as pure protein, and indicates that the thickenings are comprised of material that is similar to the fibers themselves.

Thin section electron microscopy permits the monitoring of nuclear disaggregation and the decondensation process at the ultrastructural level. Chromatin during the early stages of disulfide reduction is comparatively condensed, displaying internal disaggregations of coarse fiber clumps (Fig. 12). As the reduction reaction continues, the chromatin fibers become progressively dispersed (Figs. 13 and 14), until the whole head structure is quite expanded (Fig. 15). Any nuclear vacuoles present in the early stages are no longer resolvable by the expanded stage.

Fibers sectioned longitudinally often exhibit diameters that range between 170 and 240 Å. These fibers merge to form thicker trunks, but they do not retain their separate identities since no substructure is resolved to indicate that a union has occurred. Kaye (1969) suggested that the union of branches in spermatid chromatin is accompanied by an intrinsic structural change. This change would also result in a loss of morphological identity by fiber branches as they are transformed into a single thicker structure. In the final stages of disaggregation and decondensation, the nucleus becomes highly expanded. The rupturing and detachment of the head membranes are completed at this stage, with only remnants still attached, while the chromatin fibers are no longer clumped, but uniformly dispersed.

Whole-mount electron microscopy provides an even more precise view of the chromatin fibers during the various stages of disaggregation. The completely disaggregated unit fibers have a diameter range of 140–240 Å. Very thin fibers, with diameters of approximately 100 Å, are also occasionally observed; these appear as stretched regions of the unit fibers. Thick fibers, evidently aggregates of the unit fibers, have diameters as large as 700 Å. Often these thick fibers form converging branches, as observed similarly in their counterparts seen by thin section microscopy.

High resolution microdensitometric scans reveal that the branches have a total dry mass which collectively approximates the mass of the complete trunk (Fig. 16). The diameters of the branches range between 150 and 240 Å, while the trunk diameters range between 300 and 450 Å. No infrafiber substructure can be detected, except in one rare instance when a less electron opaque area was observed paralleling the median of a thick fiber. The densitometric scans strongly indicate that the thick trunks are aggregates of the branches, and further suggest that they are probably undergoing a process of disaggregation into unit fibers. Moreover, in order to form the aggregate, the union of unit fibers is apparently accompanied by other complex changes. In any one set of branching points, the trunk diameters are 15–30% smaller than the sums of diameters of their converging branches; yet the integrated density of the trunk remains approximately the same as the sum of densities

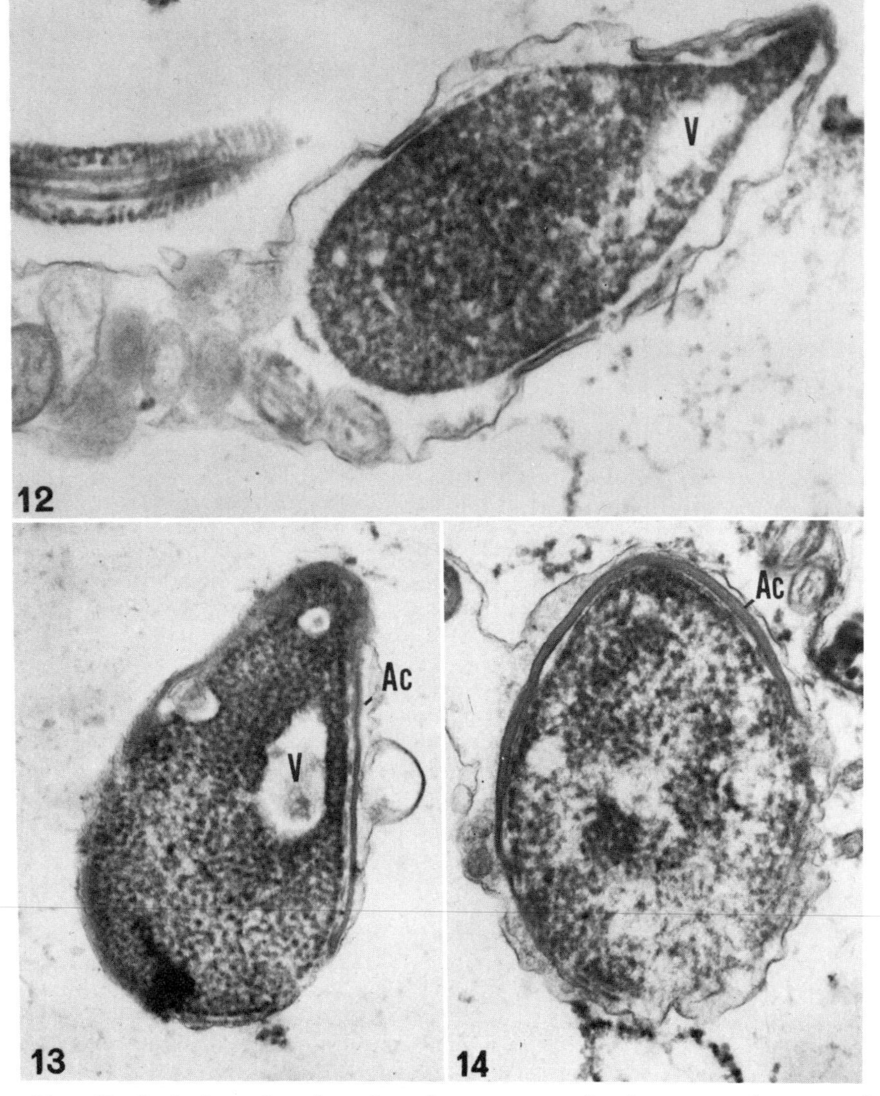

FIG. 12. Sagittal section through a human sperm head at an early stage of thioglycolate treatment. The nucleus contains thick and clumped fibrous chromatin, interrupted by an anterior vacuole (V). The head membranes of this sperm have detached early from the nucleus. (From Lung, 1972.)

FIG. 13. Sagittal section through a human sperm head in a slightly later stage of thioglycolate treatment. The chromatin fibers are at an early stage of dispersal. Nuclear vacuoles (V) are present in the anterior part of the nucleus. The acrosome (Ac) is still attached to the head, covered by a ruptured cytoplasmic membrane. (From Lung, 1972.)

FIG. 14. Frontal section through a human sperm head, showing further dispersion of chromatin. A nuclear vacuole is faintly visible. The acrosome (Ac) is still attached to the nucleus, and the ruptured cytoplasmic membrane is greatly distended. (From Lung, 1972.)

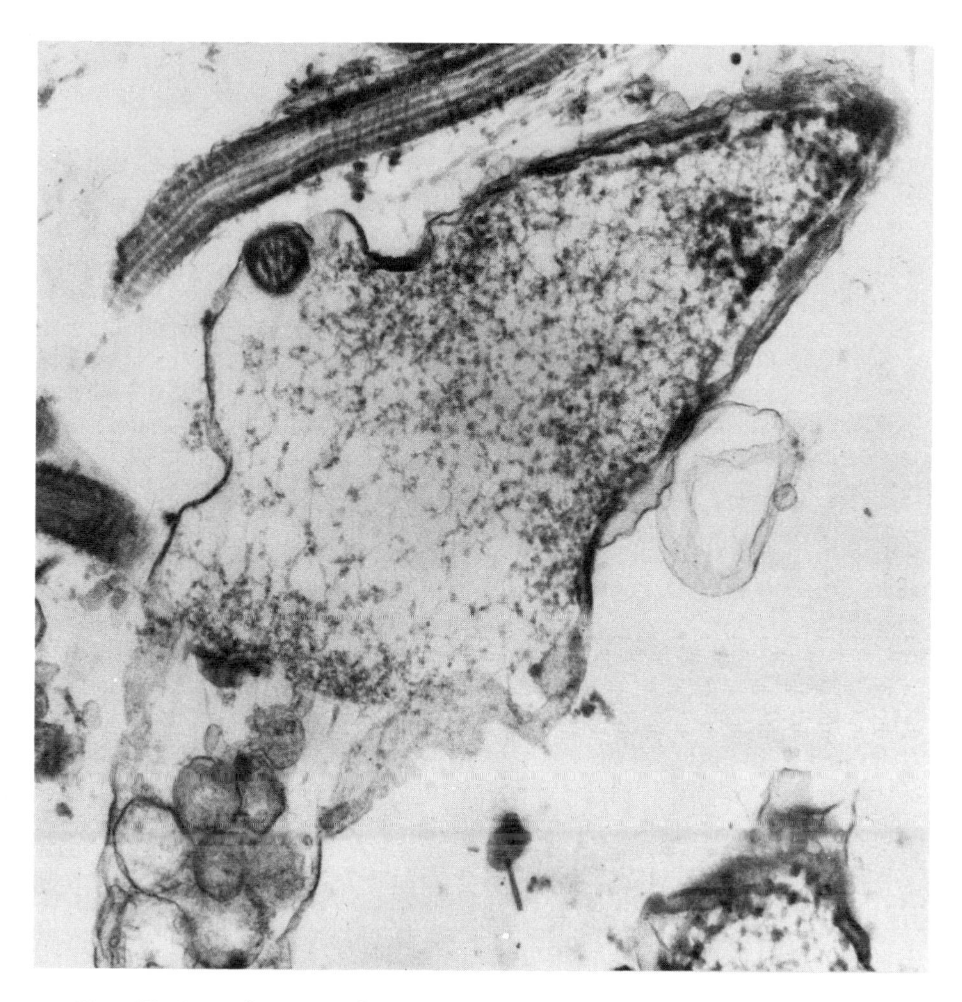

FIG. 15. Sagittal section through a human sperm head, with the nucleus in a greatly expanded state. Thin chromatin fibers are highly dispersed and undergoing a process of disaggregation. Connections may be seen between thin fibers sectioned longitudinally. The acrosome and head membranes are in a state of disruption. The cross section of a sperm tail in the upper last part of the nucleus belongs to a neighboring sperm. (From Lung, 1972).

of the branches. The side-by-side association of the unit fibers, as determined by microdensitometric scans, also suggests that the union is accompanied by intrinsic changes. At present little is known about these changes, but in mammals they seem to include at least to some extent the formation of disulfide bonds (Calvin and Bedford, 1971; Lung, 1972).

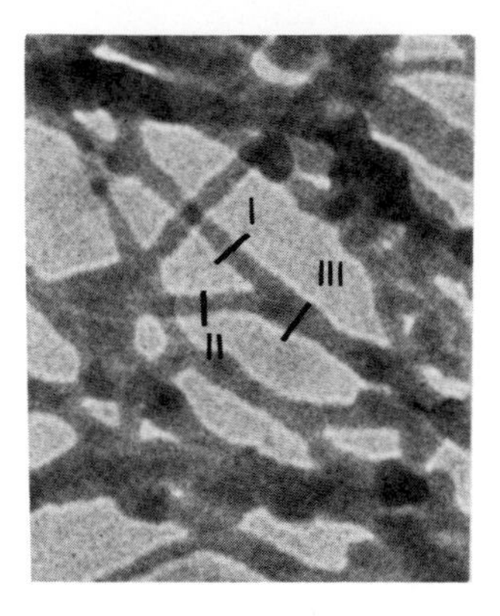

FIG. 16a. High magnification micrograph of human sperm chromatin fibers, showing branching point. High-resolution densitometric scans were performed on fiber cross sections at branching points with arms of nearly equal diameters (lines I and II) and the trunk (line III).

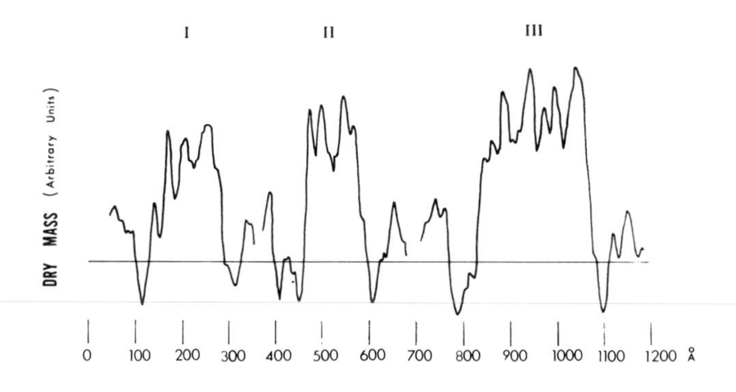

FIG. 16b. High-resolution densitometric tracings of chromatin fiber cross sections shown in Fig. 16a. Areas beneath the curves represent cross-sectional dry mass of fiber equivalents in arbitrary units. The dry mass of arm I (curve I = 478 units) and arm II (curve II = 512 units) have sums approximating the mass of the trunk (curve III = 1146 units). (Lung, 1972.)

b. Enzyme Digestion of Sperm Chromatin. A positive Schiff reaction after Feulgen staining of thioglycolate-treated mammalian sperm indicates that the DNA of the exposed nucleus resides within the chromatin fibers (Lung, 1968). Examination of chromatin prepared by the usual methods for electron microscopy, however, gives little information about the number of DNA molecules packaged within each fiber. An alternative approach is suggested via dissection of the fiber by enzymatic digestion (DuPraw, 1965).

Brief digestion of disaggregated human sperm chromatin has been performed with 100 μg of trypsin per milliliter (Worthington Biochemical, 1× crystallized) in 0.2 *M* Tris buffer at pH 8.0, warmed to 37°C. Specimens were treated for 3–5 minutes, with double controls performed by treatment with boiled enzyme and with buffer alone. After digestion, the sperm nuclei have a "thinned out" appearance, with individual fibers showing partial digestion by the trypsin (Fig. 17). However, digestion does not destroy the networklike appearance of the sperm chromatin, which is still recognizable. At high magnification, individual fibers have thinnings along their lengths (Fig. 18), producing a structure comparable to the so-called "beads on a string" effect described by DuPraw (1965) for digested chromosome fibers in honeybee embryonic cells. DuPraw's results were later confirmed by Ris (1966) in salamander erythrocytes and by Abuelo and Moore (1969) in human metaphase chromosomes. The thick portions of the digested fibers (or "beads") have diameters similar to those of undigested sperm chromatin fibers, measuring approximately 250 Å in diameter. The thin core (or "string" portion), where trypsin had digested away portions of the nucleoproteins, is resistant to further trypsin digestion; these portions have diameters measuring 20–60 Å. Some diameters approaching 80 Å have also been found, where digestion apparently was incomplete or the platinum metal seemed to have accumulated. These core diameters in human sperm chromatin are in a range consistent with those determined for interphase and metaphase chromosomes (DuPraw, 1965; Ris, 1966; Abuelo and Moore, 1969), and agree with the diameters derived from a single segment of DNA double helix.

Usually, only a single continuous, trypsin resistant "string," or core, is observed connecting the beaded portions in each chromatin fiber of human sperm. Occasionally, however, thick fiber trunks are also observed containing branches (Fig. 19). Each of the fiber branches is noticeably digested by trypsin and displays a single trypsin-resistant core, as observed for solitary, completely disaggregated fibers.

To clarify further the chemical nature of the trypsin-resistant cores, sperm chromatin fibers were first partially digested with trypsin as described previously and then treated with 10 μg of DNase I. (Worthington, beef pancreas, 1× crystallized) per milliliter in 0.003 *M* phosphate buffer with 0.0003 *M* $MgCl_2$ at pH 7.3, warmed to 37°C; digestion was for 5–15 minutes, with double controls. Comparison of fibers treated only with trypsin (Fig. 20) versus trypsin followed by DNase (Fig.

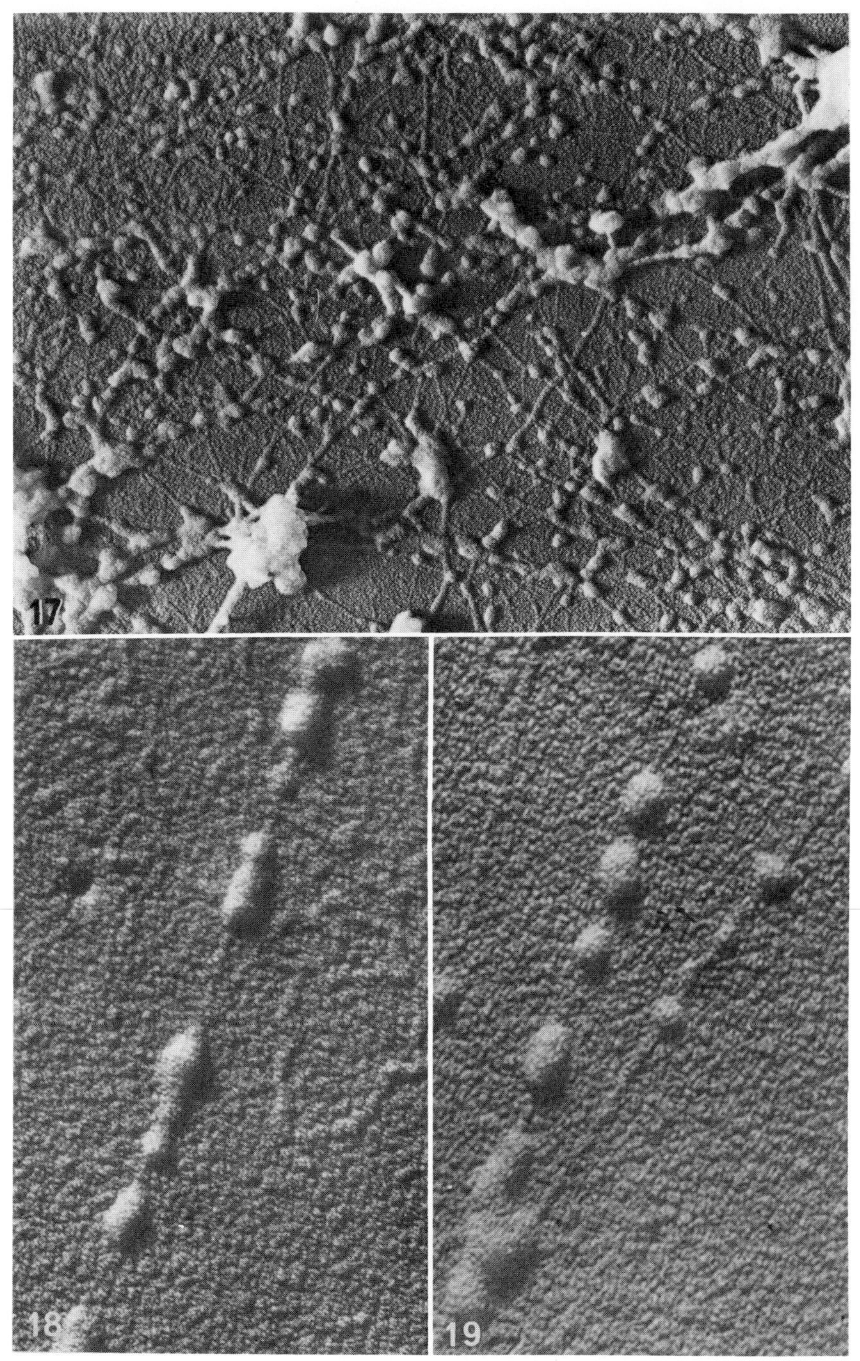

FIG. 17. Chromatin from human sperm after treatment with trypsin. Brief digestion reveals the chromatin fibers to have a "beaded" appearance, connected by a thin,

21) reveals a conspicuous absence of the trypsin-resistant cores in the DNase-digested fibers. The activity of the DNase did not produce any visible changes in the beaded portions of the fibers, as determined by diameter measurements. The sensitivity of the trypsin-resistant cores to DNase digestion strongly implies that each core is comprised of DNA, apparently arranged as a single continuous helix, which has a one-to-one relationship with the disaggregated chromatin unit fiber. The possibility that DNase merely strips DNA from the fiber, rather than digesting it, seems unlikely because only partial digestion occurs at low DNase concentrations (Abuelo and Moore, 1969).

From a technical standpoint, the action of trypsin is best observed in fibers that were attached to the support surface; those fibers extending above the surface plane appear to be less affected by the enzyme. This suggests that partial removal of the nucleoproteins destroys the original conformation of the chromatin fiber. In somatic cells, metaphase chromosomes after proteolytic digestion assume a relaxed and elongated state (Trosko and Wolff, 1965; Abuelo and Moore, 1969). Rupturing of nucleohistone intermolecular bonds by proteolytic digestion seems to cause a loss of deoxyribonucleoprotein rigidity, and destroys the original chromatin conformation (Sarka and Dounce, 1961; Fredericq, 1962). In the case of human sperm chromatin fibers, removal of the complexing proteins also appears to result in a loss of rigidity by the fibers, which subsequently collapse onto the support surface.

DuPraw (1965) reported that honeybee interphase nuclei digested by trypsin exhibit DNA that appears to have "sprung out," the chromatin fibers losing much of their original tangled appearance. The digested fibers of human sperm also have an extended appearance. The "liberated" sperm DNA between undigested fiber portions usually has a rather unraveled condition. This instant extension of the DNA limits attempts to determine its original tertiary packing configuration prior to proteolytic digestion.

c. Quantitative Electron Microscope Measurements. i. Human sperm chromatin fiber weights. Quantitative electron microscopy is an estab-

---

trypsin-resistant strand. ×59,900.

FIG. 18. High magnification view of a single chromatin fiber from human sperm digested with trypsin. The "beaded" portions of the fiber have diameters similar to those of an undigested fiber. The diameters of the trypsin-resistant connecting strand range from 30–60 Å. ×137,000.

FIG. 19. High magnification view of chromatin fibers from human sperm at a branching point after brief trypsin digestion. The branching arms of the fiber each contain a trypsin-resistent strand. ×139,100.

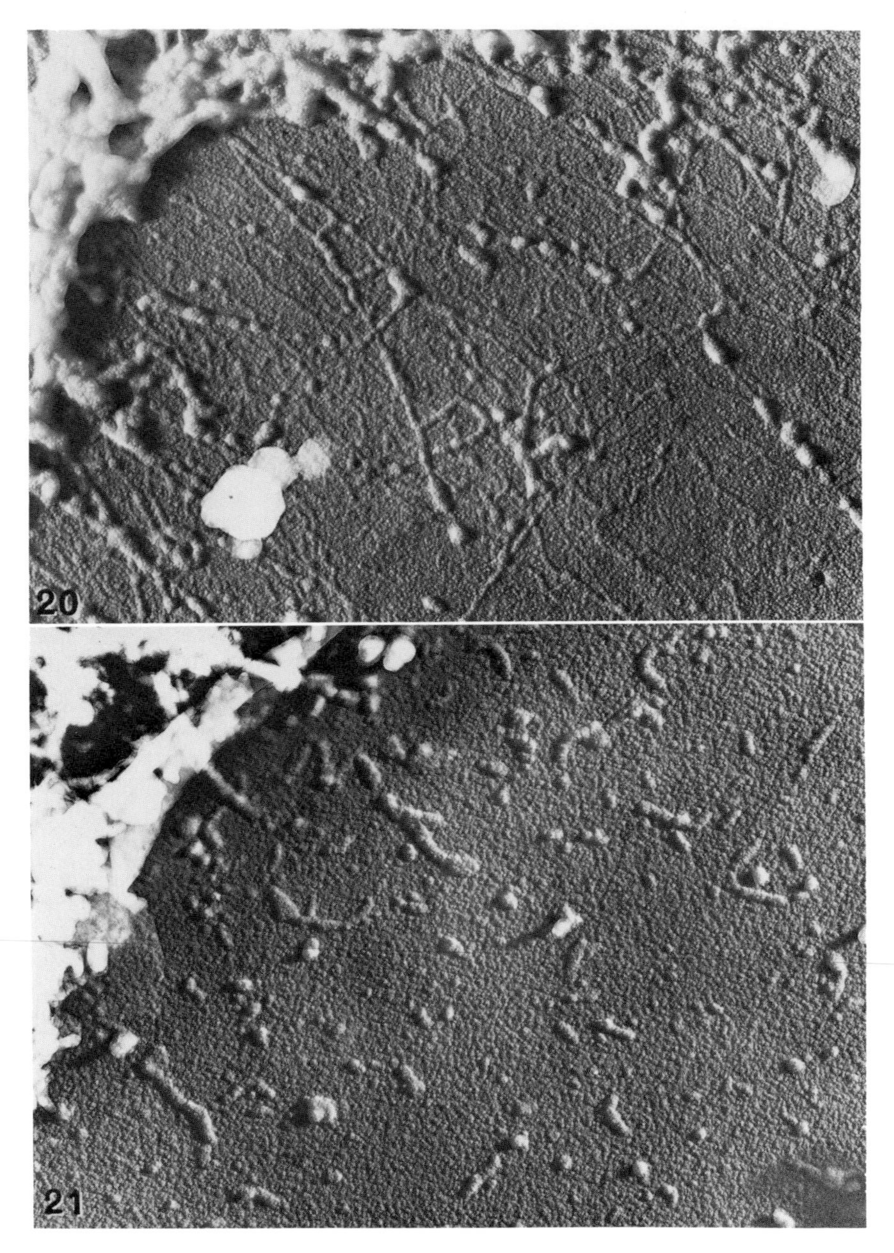

FIG. 20. Nucleus of a human sperm treated with trypsin. The chromatin fibers adhering to the grid surface show partial digestion. ×57,800.

FIG. 21. Nucleus of a human sperm treated with trypsin as in Fig. 20, but followed by digestion with DNase. The trypsin-resistant connecting strands are removed by the action of the DNase, leaving thick or "beaded" portions of the fiber not digested by trypsin. ×57,600.

lished technique for weighing biological objects with very small masses (Zeitler and Bahr, 1962; Bahr and Zeitler, 1965; Bahr, 1966). This precise method may be used to determine the dry masses of particles weighing from 10^{-11} to 10^{-18} gm on the basis of their electron-scattering capabilities when measured relative to standards of known masses and translated by densitometry (Bahr and Zeitler, 1965; Bahr, 1966). Successful application of this technique has been demonstrated in the study of human metaphase chromosomes (DuPraw and Bahr, 1969; Lampert *et al.*, 1969).

Human sperm chromatin fibers, decondensed and disaggregated with thioglycolate, have been measured to determine their absolute dry masses by quantitative electron microscopy. Each series of measurements were calibrated independently with polystyrene sphere standards (Fig. 22). A good linear correlation with zero intercept was obtained between

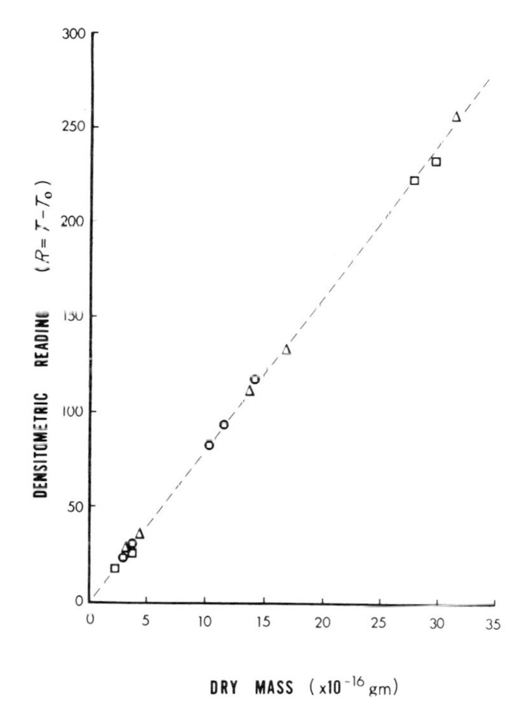

FIG. 22. A calibration curve for quantitative dry mass determinations by electron microscopy. The calculated dry mass of polystyrene sphere standards is linearly correlated with the differences in integrated transmission between the spheres and their background (densitometric reading, $R = T - T_0$). The squares, circles, and triangles represent different sphere sizes from three different calibration micrographs in a single 12-micrograph series. The average slope of the curve (broken line) is 0.122×10^{-16} gm per density unit.

the densitometric readings ($R = T - T_0$) and calculated dry masses of polystyrene spheres. In the calibration curve illustrated, polystyrene spheres of different sizes had weights ranging from 2.0 to 31.0×10^{-16} gm. The slope of the curve was 0.122×10^{-16} gm, from which the unknown fiber weights could be determined (DuPraw and Bahr, 1969; Lampert *et al.*, 1969).

Absolute dry mass was measured for sperm chromatin fibers having various diameters, but with fairly uniform widths along each axis (Figs. 23 and 24). Most of the weight measurements were limited to fibers

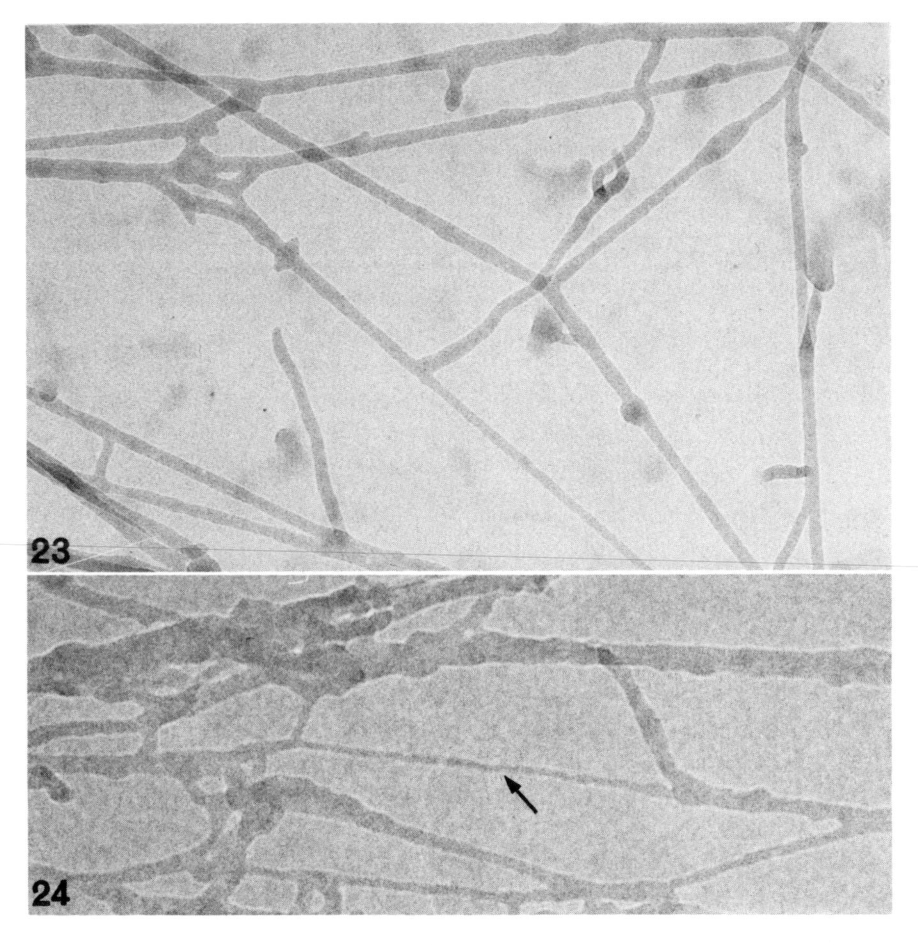

FIG. 23. Example of chromatin fibers from human sperm photographed by quantitative electron microscopy. The fibers are disaggregated, having diameters of about 200 Å, and were completely transilluminated by the electron beam. ×91,500.

FIG. 24. Chromatin from human sperm, showing a fiber that appears to have been stretched, having a diameter of about 100 Å (arrow). ×156,500.

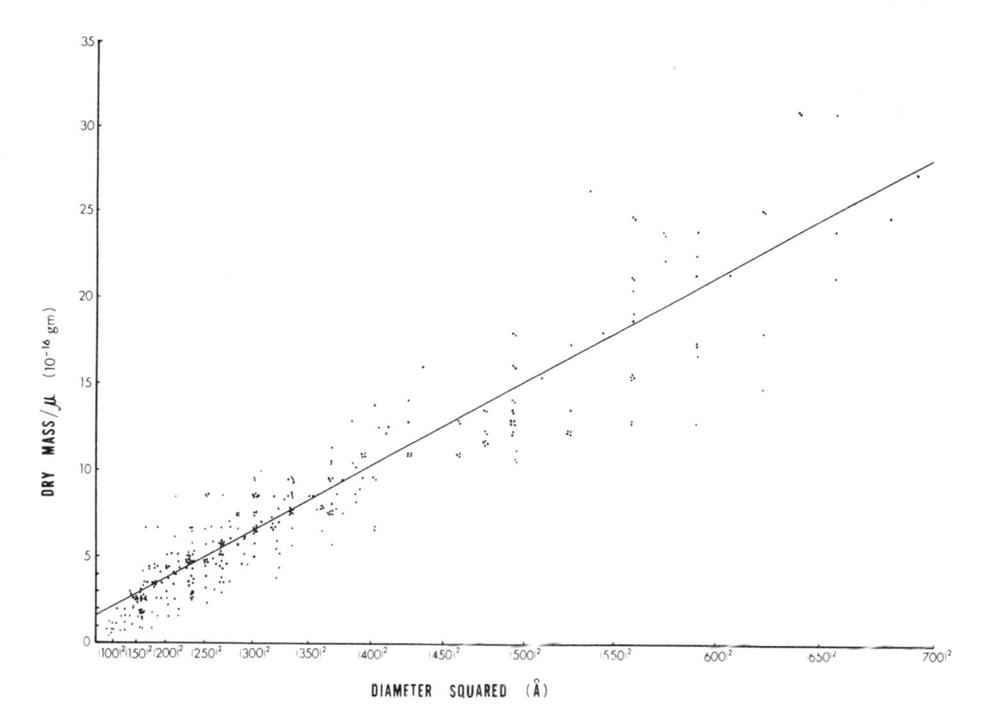

FIG. 25. Dry mass per micron of chromatin fibers from human sperm plotted against the square of fiber diameter. Quantitative measurements were performed on 330 fiber sections. The least square line (solid line) has the formula, $Y = 1.59 + 0.54X$.

with widths ranging from 100 to 700 Å, and approximately 80% of the weighed fibers had widths ranging from 100 to 450 Å. The chromatin fibers were categorized into four diameter groups: (1) thin fibers, observed where stretching seemed to have occurred, having diameters of approximately 100 Å; (2) a single, completely disaggregated unit fiber, having a diameter range of 140–250 Å; (3) a simple aggregate of two unit fibers, having a diameter range of 300–450 Å; and (4) very thick fibers representing highly aggregated groups (greater than two fibers), having diameters as large as 700 Å. The measured segment length of the fibers ranged between 500 and 1200 Å, depending on the exact plate magnification. All the fiber dry masses were normalized to a hypothetical unit length of 1μ and were plotted against the squares of each fiber's measured diameter (Fig. 25).

As seen in Fig. 25, fiber dry mass per micron increases linearly with the square of fiber diameter. The distribution of dry weights is based on 330 fiber measurements, in which dry mass ranged from 3.8×10^{-17} to 3.1×10^{-15} gm per micron. A line fitted to all the points by the method of least squares has the formula $Y = 1.59 + 0.54X$, where Y is the dry mass per micron in units of 10^{-16} gm/μ and X is the square of fiber diameter in angstrom units (diameter $\times 10^{-2}$ Å)^2. This least square line is a regression line, by which a fiber's dry mass per micron length may be predicted on the basis of its width. The regression line has a correlation coefficient of 0.92, while the average deviation in mass per micron for fibers with diameters up to 450 Å is 1.1 units; for fibers 450–700 Å, the value is 3.5 units.

As indicated by the deviation from expected Y values and the correlation coefficient, there is a degree of random variation in mass per micron for fibers of similar widths. This variation is particularly large for fibers having diameters greater than 450 Å, and for these fibers the degree of accuracy of the fitted line is less reliable. Although the measurements were performed on fibers selected for fairly uniform diameters, slight changes in the widths of some fibers could be detected. In addition, any departure of fiber geometry from a uniform cylinder would also increase random variability; a difference of 25% in width versus height would result in a variation of mass per micron by a factor of $(0.55 \times 10^{-2} \times \text{diameter})^2$. Thus, a 200 Å fiber could be expected to vary as much as 1.1 units, and a 500 Å fiber might have a departure value as great as 7.1 units.

As determined from the least-square formula, a single unit fiber having a median diameter of 200 Å should have a dry weight of 3.75×10^{-16} gm/μ. A simple aggregate of two unit fibers, having a median diameter of 375 Å, would have a dry weight of 9.2×10^{-16} gm/μ. This value, although somewhat larger than twice that of a single unit fiber, seems to confirm the relative dry mass determinations carried out on branching fiber equivalents by scanning densitometry. A higher order of fiber aggregation into thicker bundles is suggested on the basis of ultrastructural evidence as well as by dry mass determinations. The thickest fibers, reaching approximately 700 Å in diameter, have a dry mass per micron requiring units of 10^{-15} gm/μ. Thin fibers, having diameters of approximately 100 Å, have a dry mass per micron in the order of 10^{-17} gm/μ. These fibers do not seem to represent a distinct group, but rather appear as regions in which a breakdown in the tertiary conformation of the unit fiber has occurred (Fig. 24). A similar loss in tertiary conformation has been suggested to occur in the 230 Å fibers of somatic chromosomes (Gall, 1966; DuPraw and Bahr, 1969), and subsequently this was shown to be induced artificially by treatment with 1 *M* NaCl (Lampert and Lampert, 1970).

Dry mass determinations have recently been carried out on the interphase and metaphase fibers of human chromosomes (DuPraw and Bahr,

1969; Lampert *et al.*, 1969). The regression line for human interphase and metaphase chromosome fibers has a formula of $Y = 2.0 + 0.78X$ (DuPraw and Bahr, 1969); for Burkitt's lymphoma chromosome fibers, the formula is $Y = 3.0 + 0.69X$ (Lampert *et al.*, 1969). The regression formula determined for chromatin fibers from human sperm differs somewhat from these, but this is not too surprising in view of the massive replacement of somatic histones by a less complex nucleoprotein during spermiogenesis (Coelingh *et al.*, 1969; Bloch, 1969; Vaughn, 1966; Gledhill *et al.*, 1966; Monesi, 1965). Other important differences are that human sperm fibers are collectively narrower in diameter, and for fibers of similar widths their weight is less than that of somatic chromosomes. Furthermore, somatic chromosomes apparently differ from sperm chromosomes in that they do not seem to condense by an aggregation of fibers into thick bundles, but apparently change metabolic states by a shortening or contraction of the fiber (DuPraw and Bahr, 1969).

ii. Human sperm nuclear weights. Human sperm, decapitated by sonication and treated briefly with thioglycolate, exhibit nuclei in a partially condensed state which is suitable for dry mass determinations by quantitative electron microscopy (Fig. 26). These nuclei, approximately twice

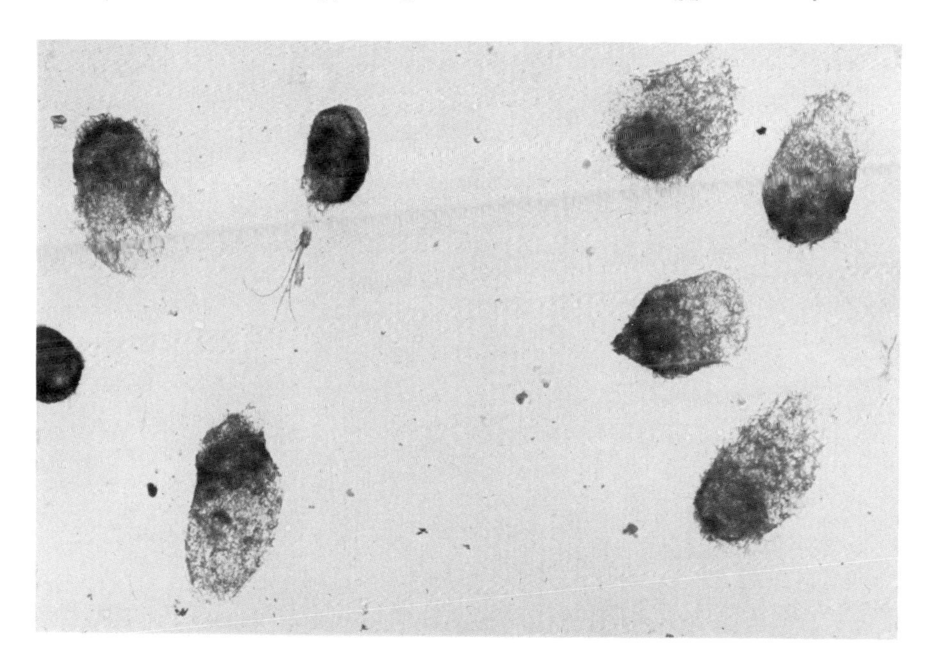

FIG. 26. Electron micrograph of human sperm nuclei photographed by quantitative electron microscopy. Examples of six nuclei are seen, which have dry masses between 5.7 and 6.7×10^{-12} gm. $\times 2500$.

their original diameter, appear as a skein or network of fibers which for the most part are devoid of membranes (except for some occasional remnants). Measurements were performed by quantitative electron microscopy on such nuclei, which were completely detached from their tails and had no contaminating membranes.

The dry masses of 295 nuclei were found to range from 4.3 to 8.3×10^{-12} gm. The mean dry mass of such sperm nuclei was 6.5×10^{-12} gm, with a standard deviation of $\pm$ 0.7 unit (Fig. 27). On the basis of interference microscopy, Leuchtenberger and Leuchtenberger (1958) reported a mean value of 6.3×10^{-12} gm for human sperm nuclei. Consequently, the dry mass determined by quantitative electron microscopy agrees well with this value. On the other hand, a somewhat lower value was reported by Sandritter *et al.* (1960a), who for reasons of geometry corrected their results by two-thirds. Without this correction, their values ranged between 6 and 7×10^{-12} gm per nucleus.

iii. Fiber length per sperm nucleus. The dry masses of human sperm nuclei are most significant when used to calculate the ratios of DNA and chromatin lengths found within each nucleus. For example, in a disaggregated nucleus composed of fibers, the mass per micron of a single unit fiber with a diameter of 200 Å (median of 140–250 Å fibers) is 3.75×10^{-16} gm/μ. From the measured nuclear weight of 6.5×10^{-12}

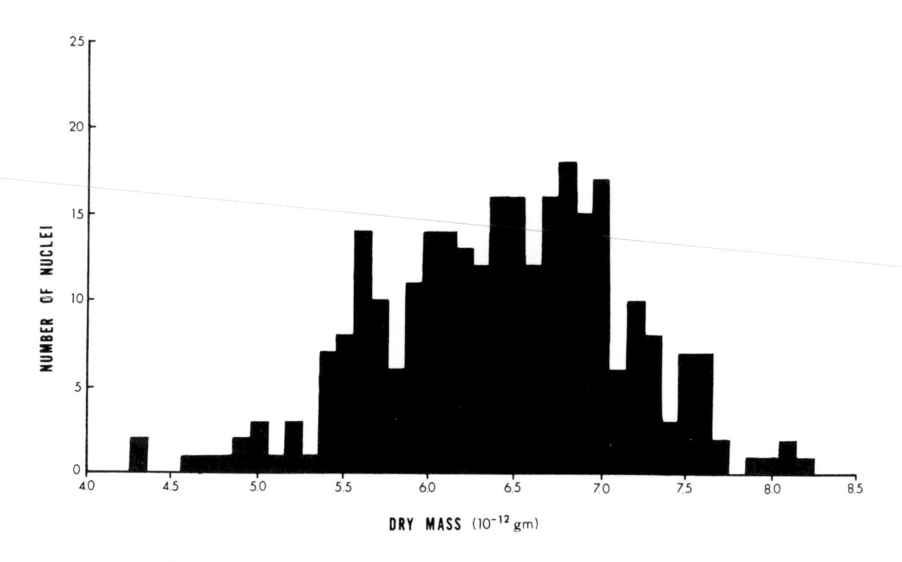

FIG. 27. A histogram illustrating the distribution of total dry mass per human sperm nucleus, as determined by quantitative electron microscopy. The total number of nuclei measured was 295. The mean dry mass of the nuclei is 6.5×10^{-12} gm, with a standard deviation of 0.7 unit.

gm, the calculated length of unit fiber in a sperm nucleus would be 1.73 cm. This value would be the theoretical maximum effective length of unit fiber found in the nucleus, irrespective of the degree of aggregation the fiber segments may undergo in the intact sperm.

As estimated by UV spectrophotometry, the amount of DNA in the nucleus of a human sperm is 3.12×10^{-12} gm (Sandritter *et al.*, 1960b); a single micron of DNA helix weighs 3.26×10^{-18} gm/μ (DuPraw and Bahr, 1969; DuPraw, 1968, 1970), which by simple division gives a total length of 95.7 cm of DNA helix to be found in each human sperm nucleus. By calculating the ratio between DNA and unit fiber lengths (95.7/1.73), a packing ratio of 55:1 is derived. Translated into simpler terms, this means that for every length of unit fiber there is on the average an equivalent of 55 lengths of DNA packaged within it. As demonstrated by enzymatic digestion, the DNA in each human sperm unit fiber appears to be a single, continuous DNA helix. This implies that the DNA in human sperm is tightly folded or coiled within the chromatin fiber.

This high sperm DNA packing ratio agrees with fiber packing ratios determined for chromosomes of human somatic cells. For example, a DNA packing ratio of 56:1 was found for human interphase chromatin (DuPraw and Bahr, 1969), and a ratio of 59:1 for Burkitt's lymphoma metaphase chromosomes (Lampert *et al.*, 1969). In addition, a much higher ratio of over 100:1 has been determined for normal lymphocytic metaphase chromosomes, suggesting that the chromosome fibers may undergo contraction during the interphase metaphase transition (DuPraw and Bahr, 1969). This contraction mechanism does not seem to be operative, however, in human sperm chromatin.

Since a single extended DNA helix has a mass per micron of 3.26×10^{-18} gm, 55 μ of DNA would have a dry mass of 1.79×10^{-16} gm; this would be approximately 48% (1.79/3.75) of the average mass per micron of a unit fiber. This value is in agreement with the 40–50% DNA generally accepted for human sperm, and supports the idea that a human sperm nucleus is composed entirely, or almost entirely, of chromatin fibers. Somatic chromosomes, by contrast, have a much lower percentage of DNA; metaphase chromosomes isolated in bulk have 15–20% DNA (Huberman and Attardi, 1966; Salzman *et al.*, 1966), while interphase chromatin has on the average 30–40% DNA (Huberman and Attardi, 1966; Salzman *et al.*, 1966; Bonner *et al.*, 1968). This implies that the DNA in sperm chromatin is much more concentrated within the fiber's architecture than that of somatic chromosomes. The increased concentration of DNA may be correlated with the highly differentiated nuclear state which exists in a sperm.

d. Human Sperm Chromosomes: A Model of Chromosome Packing. In order to summarize concisely the ultrastructural and quantitative data, it is convenient to synthesize this information into a model of chromosome and DNA packing in human sperm. Although any interpretation at present must be partly deductive, and many details are still incompletely understood, a useful framework can be generated for understanding the unique process by which sperm chromatin is differentiated into highly inactive chromosomes. It is therefore proposed that:

1. Human sperm chromatin is composed entirely or almost entirely of deoxynucleoprotein fibers that aggregate and condense into thick bundles, which may then further condense into a larger supermolecular structure. Aggregation of the deoxynucleoprotein fibers is integrally correlated with the formation of disulfide cross-links, and occurs from a networklike chromatin state.

2. Sperm DNA is distributed as a single DNA helix in each disaggregated unit fiber, and each helix constitutes over 40% of the chromatin fiber's dry mass.

3. The DNA helix is tightly folded or coiled within the chromatin fiber, with a series of homogeneous sulfur-rich, arginine-rich nucleoproteins complexed to the DNA giving rise to its tertiary configuration.

The idea that human sperm chromatin is composed of fibers was presented earlier. These fibers are observed not only in the meiotic stage (Comings and Okada, 1971; Hsu *et al.*, 1971) and in the developing spermatid (Bahr and Gledhill, 1974), but also in the fertilizing sperm as peripheral filaments emanating from a dense nucleus (Bedford, 1972; Yanagimachi and Noda, 1970c); fibers are the only consistent structure to be seen in the disaggregated nucleus (Lung, 1968, 1972). These chromatin fibers radiate in all directions and form thick bundles as they aggregate or coalesce during spermiogenesis. Aggregation and condensation are associated with a marked decrease in nuclear volume, until the fibers can no longer be distinguished as discrete units in thin sections.

The sperm chromatin or deoxynucleoprotein fiber is fundamentally a very basic nucleoprotein complexed tightly with DNA. In eutherians but not in metatherians (i.e., marsupials), the sperm nucleoprotein is a disulfide cross-linking, arginine-rich protein (Calvin and Bedford, 1971); intermolecular bridges of disulfide bonds interlock the arginine-rich proteins (Bril-Petersen and Westenbrinik, 1963; Coelingh *et al.*, 1969, 1972), and this in combination with the arginine basic groups stabilizes the DNA (Gledhill *et al.*, 1966). In addition, the disulfide linkages may also add to the inertness of the nucleus by preventing further synthesis of RNA (Hilton and Stocken, 1966).

The deoxynucleoprotein unit fibers of human sperm reflect the properties of whole sperm nuclei by having a very high percentage of DNA.

However, as in somatic cells, each fiber consists of a single DNA helix supercoiled, folded, or otherwise tightly packaged to give in human sperm a packing ratio of 55:1. This high DNA to fiber ratio presents a fundamental question as to the manner in which DNA may be folded or coiled within the chromatin fiber. In human sperm, present data most closely fit the tertiary supercoil DNA configuration (supercoiled-coil) suggested by DuPraw and Bahr (1969). Tight packing of DNA in chromosome fibers by means of supercoiling has been suggested from volumetric estimates (DuPraw, 1965), by biophysical calculations (Maestre and Kilkson, 1965), from hydrodynamic behavior (Ohba, 1966a), by X-ray diffraction studies (Pardon *et al.*, 1967; Bram and Ris, 1971), on theoretical grounds (Fong, 1967), by dry mass measurements (DuPraw and Bahr, 1969; DuPraw, 1968, 1970; Bahr, 1970; Lampert and Lampert, 1970), and by model construction (Crick, 1971). In the chromatin fibers of human sperm, in order to pack 55 equivalent lengths of DNA in each unit length of chromatin fiber, the deoxynucleoprotein fiber must by necessity contain two orders of packing or supercoiling. The first supercoil of DNA would be found in the stretched regions of the unit fibers, having a diameter of about 100 Å, as suggested by X-ray diffraction (Pardon *et al.*, 1967; Bram and Ris, 1971) and NaCl treatment (Lampert and Lampert, 1970). The 100 Å fiber would again be supercoiled to form the 200 Å unit fiber, giving sufficient folding to concentrate the DNA in a unit length of chromatin fiber and fulfill the requirement for a 55:1 packing ratio. This configuration of tertiary DNA supercoiling has the advantage of accounting for the high DNA percentage found in human sperm. Simpler configurations do not comply well with this requirement for concentrating the sperm DNA. Neither a single linear configuration of DNA, nor a single order of supercoiling, concentrate enough DNA to be satisfactorily fitted into these chromatin fibers.

The complexing of arginine-rich proteins with DNA in sperm chromosomes is likely to play a role in supercoiling. Not only have arginine-rich proteins in histones been implicated as a protein fraction involved in DNA supercoiling (Ohba, 1966a), but nucleohistone also increases the thermal stability of deoxynucleoproteins (Ohba, 1966b; Fredericq and Houssier, 1967), and this has also been demonstrated to occur in bull sperm (Ringertz *et al.*, 1970). The formation of disulfide cross-links in mammalian sperm may be an additional mechanism for the enhancement or stabilization of supercoiling.

Variations in DNA packing configuration between human sperm and somatic cell chromosomes are implied in calculations of the tertiary supercoiling. The higher sperm DNA percentage and the thinner unit fiber widths imply a more compact DNA configuration within each fiber.

Partial breakdown of tertiary supercoiling of DNA in sperm would not be inconsistent with the model. One example might be the nodelike thickenings occurring along the fiber axis. Owing to limitations inherent in electron microscopy, direct visualization of the tertiary DNA configuration is not currently possible. Whether intrinsic changes occur in the sperm DNA tertiary configuration as the chromatin unit fibers aggregate into larger units, or whether such changes are confined only to the nucleoprotein portions, remains speculative.

IV. Tail

A. Neck

The *neck* is a short anterior segment of the sperm tail, which is positioned between the head and the more posterior middle piece. It is a slightly constricted segment, approximately 1 μ in length, and architecturally very similar in most mammals. Principally composing the neck is the *connecting piece,* a pleomorphic organelle which serves to connect the head to the middle piece.

Although several early investigators believed that the connecting piece is directly continuous with the outer coarse fibers of the middle piece (Fawcett, 1958; Blom and Birch-Andersen, 1965; Nicander and Bane, 1962a), more recent investigations have demonstrated that it is ontogenetically distinct and develops well before the coarse fibers make their appearance (Fawcett, 1965; Fawcett and Phillips, 1969b; Gordon, 1972). In fact, the connecting piece has a dual origin, in which the two centrioles are the loci for its synthesis or assembly; only secondarily is the connecting piece joined to the coarse fibers (Fig. 28).

In the mature sperm, the connecting piece resembles a multicolumned pagodalike structure, which for purposes of description may be separated into a posterior supporting portion consisting of several striated columns, and an anterior articular portion joined to the columns and referred to as the capitulum (Fig. 9). The striated columns are composed of fibrous protein arranged into nine longitudinal rods, which characteristically show alternating light and dark periods. In Chinese hamster the light periods are narrow, measuring about 145 Å; the dense periods consist of 10 intraperiod lines and measure about 520 Å (Fawcett and Phillips, 1969b). The nine columns attach to the outer surface of the coarse fibers at their distal ends, where they are arranged more or less in a radial fashion with diameters nearly equal to those of the coarse fibers. A short distance anterior to this junctional region, however, the nine columns are reduced in number to seven by a process of lateral fusion: two sets of columns merge on either side to form two major columns, while the remaining five minor columns remain unmodified (Fawcett, 1965). At this point, the connecting piece becomes dorsoventrally flattened and laterally expanded. The seven striated columns proceed anteriorly and then become confluent with the capitulum, which forms the stout articular head of the neck. The capitulum fits into the *implantation fossa,* situated at the base of the nucleus, where its convex articular surface is closely applied to the *basal plate,* a dense layer of material that accumulates on the outer surface of the nuclear envelope.

Internal to the capitulum is the proximal centriole, which in sperm possessing spatulated heads has its long axis oriented with the plane of flattening and is inclined 45° to 90° to the long axis of the flagellum (Fawcett and Phillips, 1969b). This juxtanuclear centriole consists of the typical nine triplet tubules surrounded by the material of the connecting piece (Figs. 7 and 28). In some species, such as man, monkey, and rabbit, the material of the striated columns interdigitates only between the triplet tubules (Zamboni and Stefanini, 1971), whereas in chinchilla, hamster, and bull sperm, there is considerable deposition of material which completely enmeshes the triplets (Fawcett and Phillips, 1969b; Saacke and Almquist, 1964b).

A distal centriole is absent in the mature sperm, but may be observed during early formation of the spermatid, immediately posterior to the proximal centriole. It is oriented parallel to the longitudinal axis of the neck, and gives rise to the 9 + 2 axial filaments of the flagellum. Late in spermiogenesis, the distal centriole undergoes a gradual transformation until it finally disappears, or at least becomes undetectable as a distinct entity (Gordon, 1972).

Because the sperm of mammals do not possess any basal body, as typically found with other cilia or flagella (see Wolfe, 1972), a problem about what constitutes the kinetic center of the flagellum has puzzled many investigators. Rikmenspoel and Van Herpen (1969), on the basis of proton and X-ray bombardment, proposed that the "centriole" is the control center of sperm flagellar motion, but did not designate either the proximal or the distal centriole. Fawcett and Phillips (1969b) suggested that the distal centriole must constitute the basal body because of its role in flagellar development, and that the neck of the mature sperm must have a kinetic center and site of origin of the flagellar wave in the absence of any typical centriole or basal body. On the other hand, Zamboni and Stefanini (1971) designated the *proximal* centriole as the control center, which they postulate is connected to the motile elements of the flagellum by a modified distal centriole.

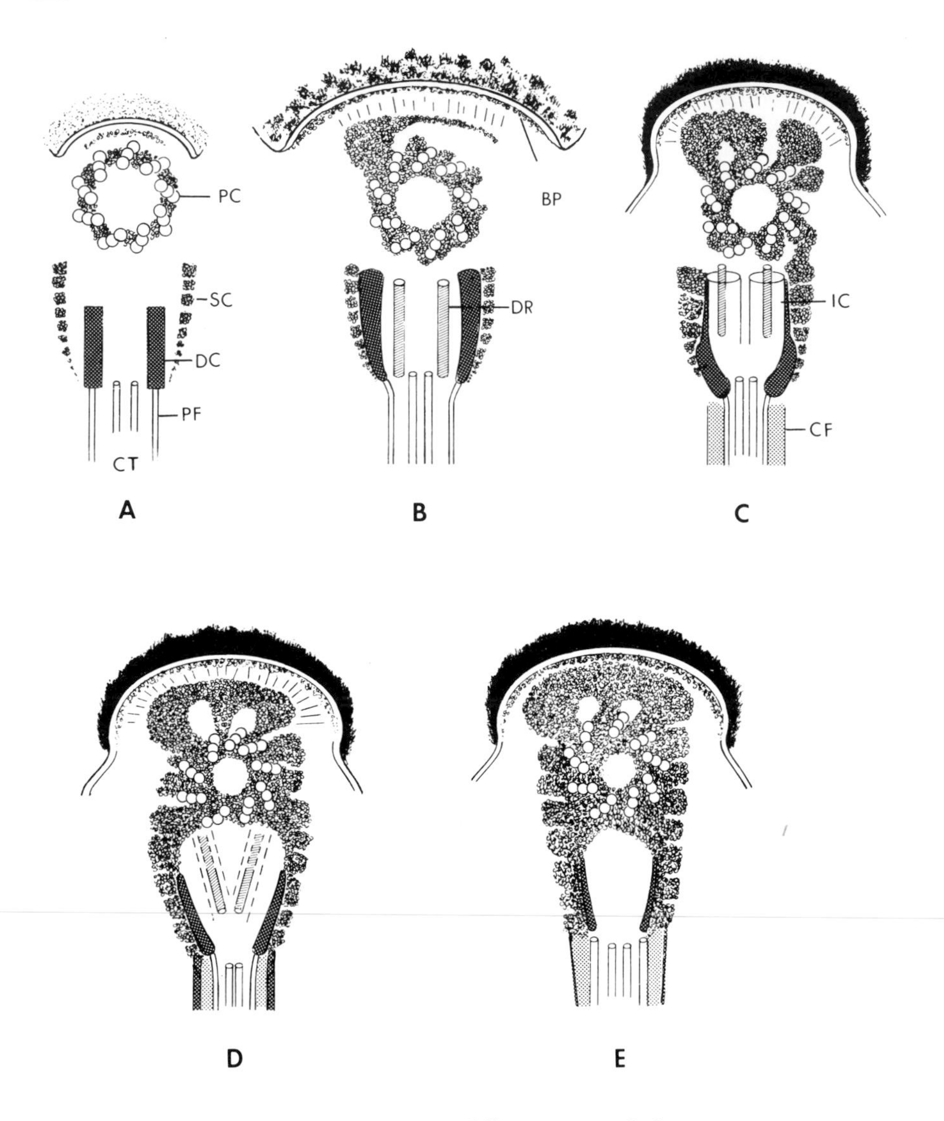

FIG. 28. A diagram of stages in the differentiation of the connecting piece, as seen in longitudinal section. PC, proximal centriole; SC, striated dense plaques; DC, distal centriole covered by dense matrix; PF, peripheral doublets; CT, central microtubules; DR, dense rods; BP, basal plate; IC, internal cylinders; CF, coarse fibers.

(A) Early stage in spermatid formation. The nucleus still displays unclumped chromatin, the proximal centriole (PC) accumulates material between the triplets of the developing connecting piece, with the anlage of the articular surface arching above. The distal centriole is enmeshed by a dense matrix (DC), with dense plaques (SC) forming at the periphery in the nearby cytoplasm. The peripheral or outer doublets (PF) of the flagellum are continuous with the distal centriole dense matrix,

This problem is further complicated by the fact that both centrioles have properties similar to basal bodies during development. While the distal centriole behaves like a typical basal body in forming the axial filament complex, the proximal centriole also forms a transient microtubular outgrowth, the *centriolar adjunct;* this consists of aberrant triplets which extend from the distal end of the proximal centriole (DeKretser, 1969; Fawcett and Phillips, 1969b). In summary, it appears that a precise organelle having the role of the kinetic center cannot yet be assigned. However, recent persuasive evidence (DeKretser, 1969; Fawcett and Phillips, 1969b; Gordon, 1972) has at least established that the connecting piece is a unique organelle, in which the two centrioles not only collaborate in microtubular synthesis, but also are sites of striated column synthesis and the assembly of several transient structures.

B. Longitudinal Fibers

1. *Axonemal Complex*

The primary mechanism of flagellar motion is a system of eleven longitudinal tubules, referred to as the *axonemal complex,* that extend the

---

and the central microtubules (CT) are forming just below the terminal end of the centriole.

(B) The nucleus aggregates into patches of chromatin clumps as the basal plate (BP) begins to thicken. The material of the proximal centriole merges clockwise with the anlage of the articular surface. The dense matrix of the distal centriole has spread slightly, fusing with the rudimentary striated columns. Internal to the centriolar dense matrix, a pair of dense rods (DR) has formed.

(C) The nucleus at this stage has become condensed. The material of the proximal centriole continues to fuse in a clockwise sequence to form the rudimentary capitulum, and lateral extensions between the triplets extend toward the anterior portion of the striated columns. The dense matrix of the distal centriole has been modified, with the formation of internal cylinders (IC) that surround the dense rods. In the flagellum, the coarse fibers (CF) are being synthesized, as outgrowths from the peripheral doublets.

(D) The material of the proximal centriole proceeds to the development of the capitulum. The lateral extensions fuse with the striated columns. The dense rods and internal cylinders of the distal centriole begin to disintegrate.

(E) The mature sperm displays a fully developed connecting piece, with the capitulum closely associated with the basal plate within the implantation fossa of the nucleus. The triplets of the proximal centriole remain unmodified, and are completely surrounded by the material of the striated columns. The only remnant of the distal centriole dense matrix is a dense internal coat covering the inner aspect of the striated columns; all vestiges of the triplets have disappeared. (Adapted and redrawn from Fawcett and Phillips, 1969b; Gordon, 1972.)

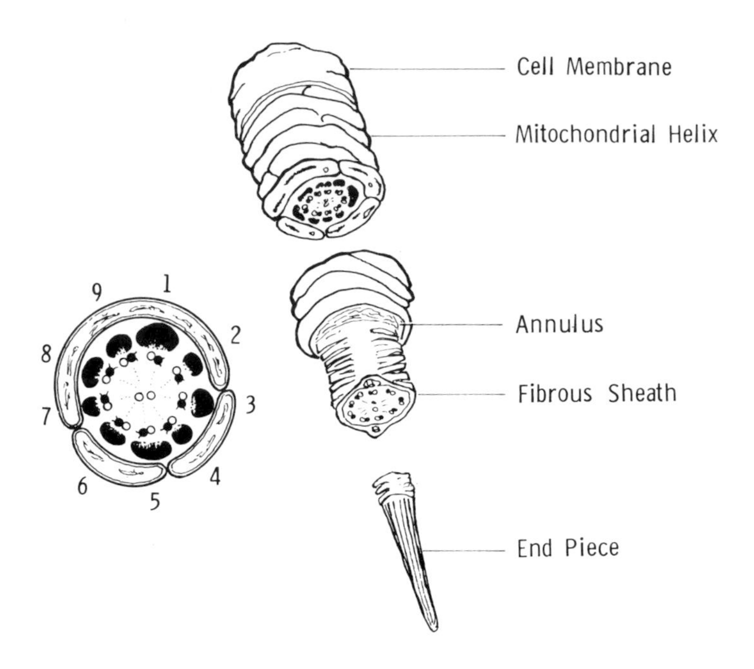

FIG. 29. Diagram summarizing the fine-structural features of the middle piece, principal piece, and end piece of a bull sperm flagellum. The cross section of the middle piece illustrates the numbering system commonly used to identify the longitudinal fibers of the tail. (Adapted and redrawn from Saacke and Almquist, 1964b.)

entire length of the flagellum (Figs. 29 and 30). This complex is arranged into the well known 9 + 2 microtubular array, and is essentially similar in structure and organization to axonemes found in other cilia and flagella (see Warner, 1972).

The axonemal complex consists of nine doublet tubules, which surround a central pair of single microtubules. Each doublet shares a common wall, and the nine are spaced uniformly, in repeated sequence, around the central axis of the flagellum. When viewed from the neck looking in the direction of the tip, one subunit of each doublet (called subfibril A) is always positioned clockwise to its counterpart (subfibril B). Two short, diverging arms protrude into the flagellar matrix from the lateral wall of subfibril A, and in favorable sections, spokelike accessory filaments may be resolved that connect the inner wall to the central pair of microtubules (Fawcett, 1962).

For most of its length, the lumen of subfibril A is completely obscured by a dense, electron opaque deposit that fills the interior and makes

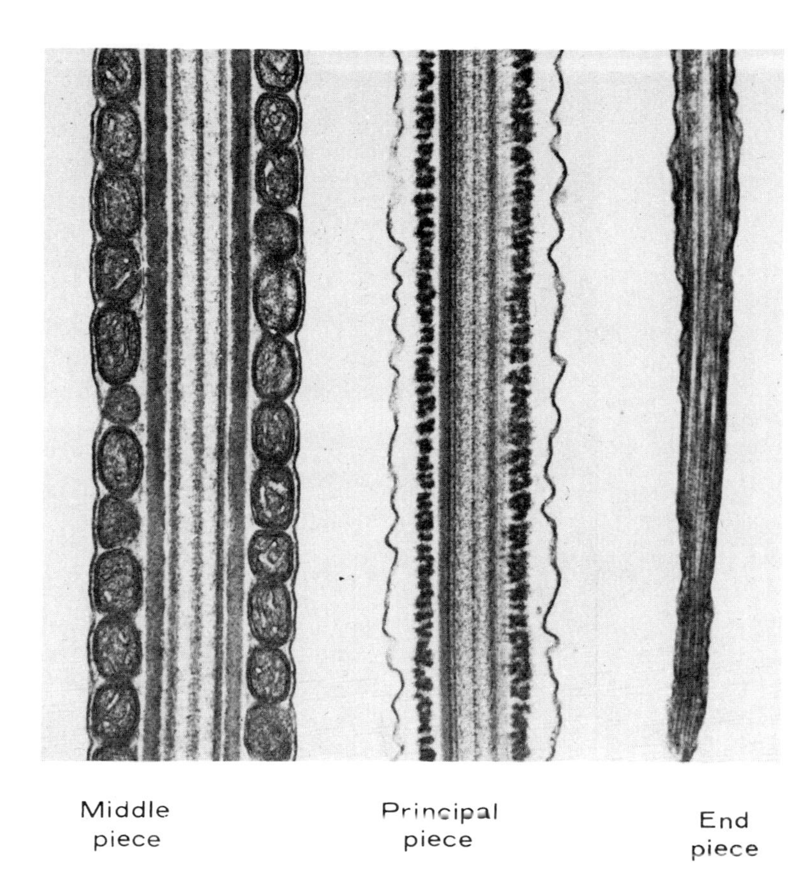

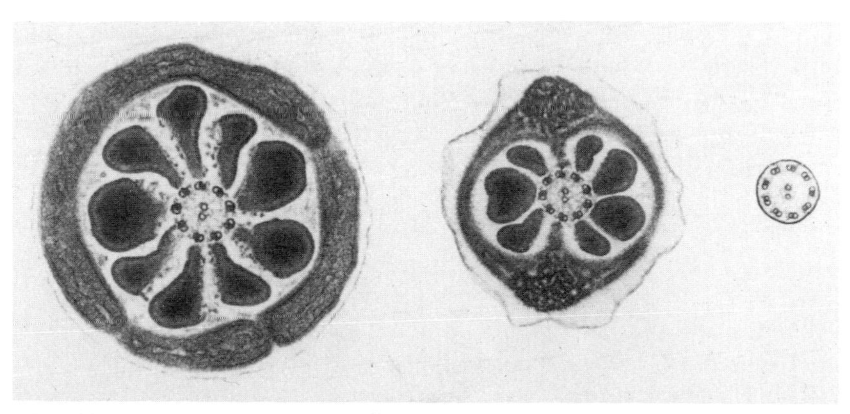

FIG. 30. Representative sections illustrating the three main segments of a mammalian sperm tail. Above, longitudinal sections through the middle piece, principal piece, and end piece of the suni, an African antelope. Below, cross sections of these main segments in Chinese hamster sperm tails. (Courtesy of Fawcett, 1970.)

subfibril A quite easy to identify. Subfibril B is centered slightly away from the radial axis, tangentially to the left of subfibril A; it differs from A by having a larger diameter, and its walls lack any projecting arms or horns. Although the lumen of subfibril B appears hollow, it is very possible that material of low electron density may be present. The central pair of microtubules is positioned centripetally to the doublets, and they consist of two separate, tubular filaments always oriented side by side. In contrast to the doublets, the walls of the central pair are never contiguous.

2. *Outer Nine Coarse Fibers*

In mammalian sperm, there is a set of nine outer coarse fibers which supplements the axonemal complex, extending through the middle piece and terminating in the anterior portion of the principal piece. These fibers, which are sometimes collectively referred to in shorthand as the $9 + 9 + 2$ fibers, lie in a ring peripheral to the axonemal doublets. In cross section, they have the striking appearance of a florally arranged cluster of commas or petals (Figs. 29 and 30). They are thickest and best developed in the proximal region of the middle piece, gradually diminishing in size as they proceed posteriorly into the tail. Structurally, the coarse fibers are dense but not entirely homogeneous (Fawcett and Ito, 1965). After differential staining, they exhibit a distinct cortical region at their outer contour, implying that they may be composite structures (Gordon and Bensch, 1968).

In accordance with the numbering system introduced by Bradfield (1955) and Afzelius (1959), which is commonly used to identify the longitudinal fibers, coarse fibers Nos. 1, 5, and 6 are the largest in cross-sectional diameter (Fig. 29). Fiber 9 varies from intermediate to large in size, and the remaining five fibers are the smallest of the group. As a rule the large fibers extend the farthest into the principal piece, while the smaller fibers are the first to terminate (Telkka *et al.*, 1961). Among the nine coarse fibers, Nos. 3 and 8 are somewhat unusual relative to the others; instead of ending blindly in the matrix of the principal piece, they become integrated with the longitudinal columns of the fibrous sheath.

Although the nine coarse fibers are characteristic of sperm produced by all mammals, they are certainly not restricted to this group. Extra flagellar fibers have also been found in certain birds, snakes, insects, and molluscs, generally in a somewhat less elaborate state of development. Emergence of the coarse fibers does not show any clear phylogenetic pattern, at least in respect to their occurrence in the animal kingdom as a whole. Instead, they are mainly limited to species that practice internal fertilization, which have a highly developed middle piece adaptively specialized for traversing a viscous medium. They are generally absent in species that fertilize externally in a more watery medium and which have a relatively simple middle piece.

Nelson (1967) suggested that the outer coarse fibers must have arisen adaptively, providing the sperm with an effective propulsive system to supplement the original ring of nine doublets; he postulated that their acquisition into the middle piece may be related to their adaptive value in meeting the added viscous load found in internal fertilization. Unfortunately, rather little is known about the physical and chemical nature of these intriguing fibers. Nelson (1962) and Gordon and Barrnett (1967), using cytochemical methods, demonstrated the presence of a number of enzymes (ATPase, pyrophosphatase, and succinic dehydrogenase); this suggests that the coarse fibers may participate actively as accessory elements in the movement of the flagellum. Such a possibility is further strengthened by the close relationship these fibers seem to have with the 9 + 2 microtubules during their development. Fawcett and Phillips (1970) showed that the outer nine coarse fibers form from lateral outgrowths, which arise from the walls of the axonemal doublets. Nevertheless, present evidence that the coarse fibers take an active role in flagellar motility remains largely circumstantial. This is due to the fact that it has not yet been possible to devise an experiment that directly correlates the coarse fibers with the actual physical mechanism of motility.

C. Middle Piece

The *middle piece* is a complex segment of the sperm extending from the neck to the annulus (Fig. 29). It comprises about 20–25% of the tail, and is characterized by a helically arranged sheath of mitochondria that surrounds both the inner axonemal complex and the outer nine coarse longitudinal fibers.

1. *Mitochondrial Helix*

The mitochondria of mammalian sperm are arranged into a single helix flattened around the coarse fibers in a spiral fashion, and with few exceptions aligned end to end (Figs. 29 and 30). This helical arrangement typifies the middle piece of all mammals, but there is considerable variation in the length of helix from species to species. So far only crude estimates can be made concerning the number of mitochondria present. André (1962) calculated 500–600 mitochondria in rat sperm, which possess a helix consisting of 350 gyres. Fawcett (1965) observed that about two mitochondria occupy each of the 42 gyres in guinea pig sperm, resulting in an estimated 80–85 mitochondria per sperm. The number of gyres varies from 65 to 75 in bull (Saacke and Almquist, 1964b); 115 in bat (Fawcett and Ito, 1965); 65 in boar (Nicander and Bane, 1962a); 47 in rabbit (Bedford, 1964); and 10–14 in man (Schultz-Larsen, 1958).

The mitochondria of the helix are derived originally from the cytoplasmic pool, and during late spermiogenesis (maturation phase) move to their final positions in the middle piece. Each mitochondrion undergoes marked elongation as members of the group fuse with one another. It is clear, however, that not all the mitochondria take part in helix formation, since a number remain in the residual bodies even after the sperm is released (DeKretser, 1969). Prior to helix morphogenesis, the mitochondria also undergo considerable internal reorganization. In many species, the cristae reorient toward the peripheral margins, with their compartments directed parallel instead of projecting perpendicularly to the mitochondrial long axis, as commonly found in somatic cells.

André (1962), who described in detail these cristae modifications in rat sperm, also observed that extensive swelling may occur within the cristal intermembraneous spaces; the swellings develop into large clear areas, referred to as a "pseudomatrix." The matrical or intercristal spaces, on the other hand, become condensed and greatly decrease in volume; this results in a reversal of the internal density relationships usually observed in tissue mitochondria. Although such internal changes have been observed in the spermatids of many species, in bat, guinea pig, and others the mature sperm have cristae that display a typical somatic form (Fawcett and Ito, 1965; Fawcett, 1965). Such variations in different cristal conformations, although dramatic in some cases, may have particular significance in terms of the metabolic requirements of the sperm in a particular species. For example, Fawcett (1970) has suggested that the internal morphology of a mitochondrion may be correlated directly with its metabolic state. To a certain extent, this might reflect a relationship to the transformations observed in the mitochondria of liver (Hackenbrock, 1966).

2. *Annulus*

Attached to the last gyre of the mitochondrial helix is an electron dense ring, referred to as the *annulus* (Fig. 29). This arises first near the neck, apposed to the cell membrane and in close association with the chromatoid body, for which it displays a particular affinity (Fawcett *et al.*, 1970; Susi and Clermont, 1970). Near the end of development, the annulus migrates in tandem with at least a portion of the chromatoid body, until it reaches its definitive position where the two separate at the junction of the middle and principal pieces. As yet, no specific function has been established for this curious structure. Anatomically, the annulus is highly polymorphic. In bull sperm, the structure is triangular in longitudinal section, and it extends posteriorly to cover the first few ribs of the fibrous sheath (Saacke and Almquist, 1964b). In man, the annulus is quite diminutive and inconspicuous, consisting of a dense narrow band in close apposition to the cell membrane (Pedersen, 1970). Finally, in chinchilla sperm, it forms a pendulous body with a bulbous tip free of attachment (Fawcett, 1970).

D. Principal Piece

The *principal piece* is the longest propulsive segment of the sperm flagellum. It consists of a tough fibrous sheath that forms a flexible exoskeleton around the axonemal cluster of longitudinal tubules (Figs. 29 and 30).

Fibrous Sheath

This structure consists of a series of closely stacked, semicircular ribs that wrap circumferentially around the flagellum and are confluent at their extremities with two longitudinal columns. In general, these ribs are flattened bands, but frequently they form bifurcations or branches that anastomose with adjacent units. In cross sections, the ribs are seen to expand in the regions where they merge, and in many species this expansion produces an elliptical contour in the outline of the sheath (Fig. 30).

The longitudinal columns are embedded in the matrix of the ribs and are always oriented at the opposite poles of a transverse axis that intersects the central pair of microtubules. They extend the entire length of the principal piece; in cross section, they may project inward toward doublets 3 and 8, for which the columns display a particular affinity. Recently, the longitudinal columns have been reported as directly connected to doublets 3 and 8 by small dense bridges (Fawcett and Phillips, 1970). Fawcett (1970) suggested that these connections may serve as mooring devices to coordinate the bending of the fibrous sheath with the inner axonemal microtubules during flagellar beating.

In the mammalian sperm so far examined, development of the fibrous sheath proceeds in two separate stages. In most species, the first stage is characterized by the formation of longitudinal columns that arise as small deposits between the flagellar membrane and doublets 3 and 8. The second stage begins with the appearance of the rib anlagen, which form from transverse strands and later interconnect with the columns. Human sperm, however, constitute an exception departing from the normal developmental pattern. The rib precursors, apparently formed from transverse microtubules, arise well before the longitudinal columns have been laid down (DeKretser, 1969). Chemically, the fibrous sheath is believed to be composed of structural proteins. It is resistant to digestion by acids, but is readily soluble in alkalies (Bradfield, 1955). Surprisingly, the fibrous sheath has a peculiar variability in alkaline mercaptides. In bull sperm, after brief treatment the sheath is rather resistant, whereas in rabbit sperm it is quite labile, with the ribs separating from the tail before they completely dissolve (Lung, unpublished observations).

E. End Piece

The end piece is the slender apical segment of the sperm flagellum, which is located posterior to the last rib of the fibrous sheath. It is very short (less than 10 μ), and consists of the $9 + 2$ microtubular array surrounded by a minimum of cytoplasm and enclosed in the flagellar membrane (Fig. 30). Except for the terminal modifications that occur at the extreme distal end, there is little that is remarkable about the microtubules of this segment. At the terminus, the doublets become dissociated into separate singlets that are completely disarrayed into random units (Nicander and Bane, 1962a; Fawcett, 1965; Pedersen, 1970; Phillips, 1970). These tubules terminate blindly at different levels of the segment tip, and do not show any evidence of intercommunication.

Acknowledgments

I would like to express thanks to Dr. E. J. DuPraw for his encouragement and help in initiating this study, to Mrs. Sinnie Falkowski for her excellent technical illustrations, and to Dr. G. F. Bahr, Armed Forces Institute of Pathology, Washington, D.C. for providing the use of his laboratory facilities. A portion of this work is based on a Ph.D. thesis, University of Maryland.

REFERENCES

Abuelo, J. G., and Moore, D. E. (1969). *J. Cell Biol.* **41**: 73.
Afzelius, B. (1959). *J. Biophys. Biochem. Cytol.* **5**: 269.

Allison, A. C. (1967). *Sci. Amer.* **217**: 62.
Allison, A. C., and Hartree, E. F. (1970). *J. Reprod. Fert.* **21**: 501.
Anberg, A. (1957). *Acta Obstet. Gynecol. Scand.* **36** (Suppl. 2): 1.
Anderson, W. A., Weissman, A., and Ellis, R. A. (1967). *J. Cell Biol.* **32**: 11.
André, J. (1962). *J. Ultrastruct. Res., Suppl.* **3**: 1.
Austin, C. R. (1948). *Nature (London)* **162**: 63.
Austin, C. R., and Bishop, M. W. H. (1958). *Proc. Roy. Soc. Ser. B* **149**: 241.
Bahr, G. F. (1966). *In* "Introduction to Quantitative Cytochemistry" (G. L. Wied, ed.), p. 137. Academic Press, New York.
Bahr, G. F. (1970). *Exp. Cell Res.* **62**: 39.
Bahr, G. F., and Gledhill, B. L. (1974). In preparation.
Bahr, G. F., and Zeitler, E. (1965). *In* "Symposium on Quantitative Electron Microscopy" (G. F. Bahr and E. Zeitler, eds.), p. 217. Williams & Wilkins, Baltimore, Maryland.
Ballowitz, E. (1890). *Arch. Mikrosk. Anat. Entwicklungsmech.* **36**: 253.
Bane, A., and Nicander, L. (1963). *Int. J. Fert.* **8**: 865.
Barros, C., and Franklin, L. E. (1968). *J. Cell Biol.* **37**: C13.
Barros, C., Bedford, J. M., Franklin, L. E., and Austin, C. R. (1967). *J. Cell Biol.* **34**: C1.
Bearden, J., and Bendet, I. J. (1972). *J. Cell Biol.* **55**: 489.
Bedford, J. M. (1964). *J. Reprod. Fert.* **7**: 221.
Bedford, J. M. (1965). *J. Anat.* **99**: 891.
Bedford, J. M. (1967). *Amer. J. Anat.* **121**: 443.
Bedford, J. M. (1968). *Amer. J. Anat.* **123**: 328.
Bedford, J. M. (1970). *Biol. Reprod., Suppl.* **2**: 128.
Bedford, J. M. (1972). *Amer. J. Anat.* **133**: 213.
Bedford, J. M., and Nicander, L. (1971). *J. Anat.* **108**: 527.
Bendet, I. J., and Bearden, J. (1972). *J. Cell Biol.* **55**: 501.
Bishop, M. W. H., and Walton, A. (1960). *In* "Marshall's Physiology of Reproduction," 3rd Ed., Vol. 1, Part 2, p. 1. Longmans, Green, New York.
Bloch, D. P. (1969). *Genetics* **61**, Suppl. 1:94.
Bloch, D. P., and Hew, H. Y. C. (1960). *J. Biophys. Biochem. Cytol.* **8**: 69.
Blom, E., and Birch-Andersen, A. (1965). *Nord. Vet. Med.* **17**: 193.
Bonner, J., Dahmus, M. E., Fambrough, D., Huang, R. C., Marushige, K., and Tuan, D. Y. (1968). *Science* **159**: 47.
Borenfreund, E., Fitt, E., and Bendich, A. (1961). *Nature (London)* **191**: 1375.
Boyer, P. D. (1959). *In* "The Enzymes" (P. D. Boyer, H. Lardy, and K. Myrbäck, eds.), Vol. 1, p. 511. Academic Press, New York.
Bradfield, J. R. G. (1955). *Symp. Soc. Exp. Biol.* **9**: 306.
Bram, S., and Ris, H. (1971). *J. Mol. Biol.* **55**: 325.
Bril-Petersen, E., and Westenbrinik, H. G. K. (1963). *Biochim. Biophys. Acta* **76**: 152.
Brõkelmann, J. (1963). *Z. Zellforsch. Mikrosk. Anat.* **59**: 820.
Burgos, M. H., and Fawcett, D. W. (1955). *J. Biophys. Biochem. Cytol.* **1**: 287.
Burgos, M. H., Vitale-Calpe, R., and Aoki, A. (1970a). *In* "The Testis" (A. D. Johnson, W. R. Gomes, and N. L. Van Demark, eds.), Vol. 1, p. 551. Academic Press, New York.
Burgos, M. H., Sacerdote, F. L., Vitale-Calpe, R. and Bari, D. (1970b). In "The Human Testis" (E. Rosemberg and C. A. Paulsen, eds.), *Advan. Exp. Med. Biol.* Vol. 10, p. 369, Plenum, New York.

Calvin, H. I., and Bedford, J. M. (1971). *J. Reprod. Fert., Suppl.* **13**: 65.
Clermont, Y. (1967). *Arch. Anat. Microsc. Morphol. Exp.* **56**, Suppl. 3–4:7.
Coelingh, J. P., Rozijn, T. H., and Monfoort, C. H. (1969). *Biochim. Biophys. Acta* **188**: 353.
Coelingh, J. P., Monfoort, C. H., Rozign, T. H., Gevers Leuven, J. A., Schiphof, R., Steyn-Parvé, E. P., Braunitzer, G., Schrank, B., and Ruhfus, A. (1972). *Biochim. Biophys. Acta* **285**: 1.
Comings, D. E., and Okada, T. A. (1971). *Exp. Cell Res.* **65**: 99.
Comings, D. E., and Okada, T. A. (1972a). *Advan. Cell Mol. Biol.* **2**: 310.
Comings, D. E., and Okada, T. A. (1972b). *J .Ultrastruct .Res.* **39**: 15.
Courot, M., Hochereau-de-Reviers, M. T., and Ortavant, R. (1970). *In* "The Testis" (A. D. Johnson, W. R. Gomes, and N. L. Van Demark, eds.), Vol. 1, p. 339. Academic Press, New York.
Crick, F. (1971). *Nature (London)* **234**: 25.
Czermak, J. N. (1879). "Gesammelte Schriften." Engelmann, Leipzig.
Dallum, R. D., and Thomas, L. E. (1953). *Biochim. Biophys. Acta* **11**: 79.
Dan, J. C. (1967). *In* "Fertilization: Comparative Morphology, Biochemistry and Immunology" (C. B. Metz and A. Monroy, eds.), Vol. 1, p. 237. Academic Press, New York.
Darzynkiewicz, Z., Gledhill, B. L., and Ringertz, N. R. (1969). *Exp. Cell Res.* **58**: 435.
Das, C. C., Kaufmann, B. P., and Gay, H. (1964). *J. Cell Biol.* **23**: 423.
DeKretser, D. M. (1969). *Z. Zellforsch. Mikrosk. Anat.* **98**: 477.
Dietert, S. E. (1966). *J. Morphol.* **120**: 317.
DuPraw, E. J. (1965). *Proc. Nat. Acad. Sci. U.S.* **53**: 161.
DuPraw, E. J. (1968). "Cell and Molecular Biology." Academic Press, New York.
DuPraw, E. J. (1970). "DNA and Chromosomes." Holt, New York.
DuPraw, E. J., and Bahr, G. F. (1969). *Acta Cytol.* **13**: 188.
Fawcett, D. W. (1958). *Int. Rev. Cytol.* **7**: 195.
Fawcett, D. W. (1962). *In* "Spermatozoan Motility" (D. W. Bishop, ed.), p. 147. Amer. Ass. Advan. Sci., Washington, D.C.
Fawcett, D. W. (1965). *Z. Zellforsch. Mikrosk. Anat.* **67**: 279.
Fawcett, D. W. (1970). *Biol. Reprod., Suppl.* **2**: 90.
Fawcett, D. W., and Burgos, M. H. (1956). *Ciba Found. Colloq. Ageing* **2**: 86.
Fawcett, D. W., and Ito, S. (1965). *Amer. J. Anat.* **116**: 567.
Fawcett, D. W., and Phillips, D. M. (1969a). *J. Reprod. Fert., Suppl.* **6**: 405.
Fawcett, D. W., and Phillips, D. M. (1969b). *Anat. Rec.* **165**: 153.
Fawcett, D. W., and Phillips, D. M. (1970). *In* "Comparative Spermatology" (B. Baccetti, ed.), p. 2. Academic Press, New York.
Fawcett, D. M., Eddy, E. M., and Phillips, D. M. (1970). *Biol. Reprod.* **2**: 129.
Fong, P. (1967). *J. Theor. Biol.* **15**: 230.
Franklin, L. E. (1968). *Anat. Rec.* **161**: 149.
Franklin, L. E., and Fussell, E. N. (1972). *Biol. Reprod.* **7**: 194.
Fredericq, E. (1962). *Biochim. Biophys. Acta* **68**: 167.
Fredericq, E., and Houssier, C. (1967). *Eur. J. Biochem.* **1**: 51.
Gaddum, P., and Blandau, R. J. (1970). *Science* **170**: 749.
Gall, J. G. (1966). *Chromosoma* **20**: 221.
Gardner, P. J. (1966). *Anat. Rec.* **148**: 284.
Gatenby, J. B., and Beams, H. W. (1935). *Quart. J. Microsc. Sci.* **78**: 1.
Gatenby, J. B., and Wigoder, S. B. (1929). *Proc. Roy. Soc., Ser. B* **104**: 471.

Gledhill, B. L. (1971). *J. Reprod. Fert., Suppl.* **13**: 77.
Gledhill, B. L., Gledhill, M. P., Rigler, R., and Ringertz, N. R. (1966). *Exp. Cell Res.* **41**: 652.
Gordon, M. (1969). *J. Reprod. Fert.* **19**: 367.
Gordon, M. (1972). *Anat. Rec.* **173**: 45.
Gordon, M., and Barrnett, R. J. (1967). *Exp. Cell Res.* **48**: 395.
Gordon, M., and Bensch, K. G. (1968). *J. Ultrastruct. Res.* **24**: 33.
Hackenbrock, C. R. (1966). *J. Cell Biol.* **30**: 269.
Hadek, R. (1963). *J. Ultrastruct. Res.* **9**: 110.
Hadek, R. (1969). *J. Ultrastruct. Res.* **27**: 396.
Hancock, J. L. (1957). *J. Roy. Microsc. Soc.* **76**: 84.
Hancock, J. L., and Trevan, D. J. (1957). *J. Roy. Microsc. Soc.* **76**: 77.
Hartree, E. F., and Srivastava, P. N. (1965). *J. Reprod. Fert.* **9**: 47.
Hendricks, D. M., and Mayer, D. T. (1965a). *Exp. Cell Res.* **40**: 402.
Hendricks, D. M., and Mayer, D. T. (1965b). *Proc. Soc. Exp. Biol. Med.* **119**: 769.
Hilton, J., and Stocken, L. A. (1966). *Biochem. J.* **100**: 21C.
Hopsu, V. K., and Arstila, A. U. (1965). *Z. Zellforsch. Mikrosk. Anat.* **65**: 562.
Horstmann, E. (1961). *Z. Zellforsch. Mikrosk. Anat.* **54**: 68.
Hsu, T. C., Cooper, J. E. K., Mace, M. L., and Brinkley, B. R. (1971). *Chromosoma* **34**: 73.
Huberman, J., and Attardi, C. (1966). *J. Cell Biol.* **31**: 95.
Kaye, J. S. (1969). *In* "Handbook of Molecular Biology" (A. Lima-deFaria, ed.), p. 361. North-Holland Publ., Amsterdam.
Kessel, R. G. (1966). *J. Ultrastruct. Res.* **16**: 293.
Koehler, J. K. (1966). *J. Ultrastruct. Res.* **16**: 359.
Koehler, J. K. (1970). *J. Ultrastruct. Res.* **33**: 598.
Koehler, J. K. (1972). *J. Ultrastruct. Res.* **39**: 520.
Kopečny, V. (1970). *Z. Zellforsch. Mikrosk. Anat.* **109**: 414.
Lalli, M. L. (1971). *Anat. Rec.* **169**: 362.
Lampert, F., and Lampert, P. (1970). *Humangenetik* **11**: 9.
Lampert, F., Bahr, G. F., and DuPraw, E. J. (1969). *Cancer* **24**: 367.
Leblond, C. P. and Clermont, Y. (1952). *Amer. J. Anat.* **90**: 167.
Leuchtenberger, C., and Leuchtenberger, R. (1958). *Hoppe-Seyler's Z. Physiol. Chem.* **313**: 130.
Leuchtenberger, C., Leuchtenberger, R., Vendrely, C., and Vendrely, R. (1952). *Exp. Cell Res.* **3**: 240.
Leuchtenberger, C., Leuchtenberger, R., Schrader, F., and Weir, D. R. (1956). *Lab. Invest.* **5**: 422.
Loir, M. M. (1970). *C. R. Acad. Sci., Ser. D* **271**: 1634.
Lung, B. (1968). *J. Ultrastruct. Res.* **22**: 485.
Lung, B. (1972). *J. Cell Biol.* **52**: 179.
Lung, B., and Bahr, G. F. (1972). *J. Reprod. Fert.* **31**: 317.
Maestre, J. F., and Kilkson, R. (1965). *Biophys. J.* **5**: 275.
Metz, C. B. (1967). *In* "Fertilization: Comparative Morphology, Biochemistry and Immunology" (C. B. Metz and A. Monroy, eds.), Vol. 1, p. 163. Academic Press, New York.
Meves, F. (1899). *Arch. Mikrosk. Anat.* **54**: 329.
Meves, F. (1901). *Anat. Hefte, Abt.* 2 **11**: 437.
Monesi, V. (1964). *Exp. Cell Res.* **36**: 683.

Monesi, V. (1965). *Exp. Cell Res.* **39**: 197.
Monesi, V. (1967). *Arch. Anat. Microsc. Morphol. Exp.* **56**, Suppl. 3–4: 61.
Monesi, V. (1971). *J. Reprod. Fert., Suppl.* **13**: 1.
Moricard, R. (1961). *C. R. Soc. Biol.* **151**: 2243.
Muramatsu, M., Utakoji, T., and Sugano, T. (1968). *Exp. Cell Res.* **53**: 278.
Nelson, L. (1962). *In* "Spermatozoan Motility" (D. W. Bishop, ed.), p. 171. Amer. Ass. Advan. Sci., Washington, D.C.
Nelson, L. (1967). *In* "Fertilization" (C. B. Metz and A. Monroy, eds.), Vol. 1, p. 27. Academic Press, New York.
Neutra, M., and Leblond, C. P. (1966). *J. Cell Biol.* **30**: 119.
Nicander, L. (1967). *Z. Zellforsch. Mikrosk. Anat.* **83**: 375.
Nicander, L., and Bane, A. (1962a). *Z. Zellforsch. Mikrosk. Anat.* **57**: 390.
Nicander, L., and Bane, A. (1962b). *Int. J. Fert.* **7**: 339.
Nicander, L., and Bane, A. (1966). *Z. Zellforsch. Mikrosk. Anat.* **72**: 496.
O'Donnell, J. M., Symons, D. B. A., and Wooding, F. B. P. (1970). *J. Physiol. (London)* **210**: 120P.
Ohba, Y. (1966a). *Biochim. Biophys. Acta* **123**: 76.
Ohba, Y. (1966b). *Biochim. Biophys. Acta* **123**: 84.
Onuma, H., and Nishikawa, Y. (1963). *Chikusan Shikensho Kenkyu Hokoku* **1**: 125.
Pardon, J. F., Wilkins, M. H. F., and Richards, B. M. (1967). *Nature (London)* **215**: 508.
Pedersen, H. (1969). *Z. Zellforsch. Mikrosk. Anat.* **94**: 542.
Pedersen, H. (1970). *In* "Comparative Spermatology" (B. Baccetti, ed.), p. 133. Academic Press, New York.
Phillips, D. M. (1970). *J. Ultrastruct. Res.* **33**: 281.
Piko, L. (1969). *In* "Fertilization" (C. B. Metz and A. Monroy, eds.), Vol. 2, p. 325. Academic Press, New York.
Plattner, H. (1971). *J. Submicrosc. Cytol.* **3**: 19.
Ploen, L. (1971). *Z. Zellforsch. Mikrosk. Anat.* **115**: 553.
Polakoski, K. L., Zaneveld, L. J. D., and Williams, W. L. (1972). *Biol. Reprod.* **6**: 23.
Postwald, H. E. (1967). *Z. Zellforsch. Mikrosk. Anat.* **83**: 231.
Rahlmann, D. F. (1961). *J. Dairy Sci.* **44**: 916.
Rattner, J. B., and Brinkley, B. R. (1971). *J. Ultrastruct. Res.* **36**: 1.
Retzius, G. (1909a). *Biol. Untersuch.* **14**: 163.
Retzius, G. (1909b). *Biol. Untersuch.* **14**: 205.
Rikmenspoel, R., and Van Herpen, G. (1969). *Biophys. J.* **9**: 833.
Ringertz, N. R., Gledhill, B. L., and Darzynkiewicz, E. (1970). *Exp. Cell Res.* **62**: 204.
Ris, H. (1966). *Proc. Roy. Soc., Ser. B* **164**: 246.
Saacke, R. G., and Almquist, J. O. (1964b). *Amer. J. Anat.* **115**: 163.
Saacke, R. G., and Almquist, J. O. (1964b). *Amer. J Anat.* **115**: 163.
Salzman, N., Moore, D., and Mendelsohn, J. (1966). *Proc. Nat. Acad. Sci. U.S.* **56**: 1449.
Sandoz, D. (1970). *J. Microsc. (Paris)* **9**: 535.
Sandritter, W., Muller, D., and Genecke, O. (1960a). *Acta Histochem.* **10**: 139.
Sandritter, W., Schiemer, H. G., and Uhlig, H. (1960b). *Acta Histochem.* **10**: 155.

Sapsford, C. S., and Rae, C. A. (1969). *Aust. J. Zool.* **17**: 415.
Sapsford, C. S., Rae, C. A., and Cleland, K. W. (1967). *Aust. J. Zool.* **15**: 881.
Sapsford, C. S., Rae, C. A., and Cleland, K. W. (1969). *Aust. J. Zool.* **17**: 195.
Sarka, N. K., and Dounce, A. L. (1961). *Arch. Biochem. Biophys.* **92**: 321.
Schiffer, M., and Edmundson, A. B. (1967). *Biophys. J.* **7**: 121.
Schultz-Larsen, J. (1958). *Acta Pathol. Microbiol. Scand., Suppl.* **128**: 1.
Schweigger-Seidel, F. (1865). *Arch. Mikrosk. Anat.* **1**: 309.
Seiguer, A. C., and Castro, A. E. (1972). *Biol. Reprod.* **7**: 31.
Srivastava, P. N., Adams, C. E., and Hartree, E. F. (1965). *J. Reprod. Fert.* **10**: 61.
Srivastava, P. N., Zaneveld, L. J. D., and Williams, W. L. (1970). *Biochem. Biophys. Res. Commun.* **39**: 575.
Stambaugh, R., and Buckley, J. (1968). *Science* **161**: 585.
Stambaugh, R., and Buckley, J. (1969). *J. Reprod. Fert.* **19**: 423.
Stambaugh, R., and Buckley, J. (1970). *Biol. Reprod.* **3**: 275.
Stanley, H. P. (1966). *Anat. Rec.* **154**: 426.
Stanley, H. P. (1969). *J. Ultrastruct. Res.* **27**: 230.
Stefanini, M., DeMartino, C., and Zamboni, L. (1967). *Nature (London)* **216**: 173.
Sud, B. N. (1961a). *Quart. J. Microsc. Sci.* **102**: 273.
Sud, B. N. (1961b). *Quart J. Microsc. Sci.* **102**: 495.
Susi, F. R., and Clermont, Y. (1970). *Amer. J Anat.* **129**: 177.
Susi, F. R., Leblond, C. P., and Clermont, Y. (1971). *Amer. J. Anat.* **130**: 251.
Telkka, A., Fawcett, D. W., and Christensen, A. K. (1961). *Anat. Rec.* **141**: 231.
Trosko, J. E., and Wolff, S. (1965). *J. Cell Biol.* **26**: 125.
Utakoji, T. (1966). *Exp. Cell Res.* **42**: 585.
van Leeuwenhock, A. (1678). *Phil. Trans.* **12**: 1040.
Vaughn, J. C. (1966). *J. Cell Biol.* **31**: 257.
Vendrely, R., and Vendrely, C. (1948). *Experientia* **4**: 434.
von Borries, B., and Ruska, E. (1938). *Wiss. Veroeff. Siemens-Werken* **17**: 99.
von Borries, B., and Ruska, E. (1939). *Z. Wiss. Mikrosk.* **56**: 314.
Warner, F. D. (1972). *Advan. Cell Mol. Biol.* **2**: 193.
White, J. C., Leslie, I., and Davidson, J. N. (1953). *J. Pathol. Bacteriol.* **66**: 291.
Whur, P., Herscovics, A., and Leblond, C. P. (1969). *J. Cell Biol.* **43**: 289.
Wilkins, M. H. F. (1956). *Cold Spring Harbor Symp. Quant. Biol.* **21**: 75.
Wimstatt, W. A., Krutzsch, P. H., and Napolitano, L. (1966). *Amer. J. Anat.* **119**: 25.
Wolfe, J. (1972). *Advan. Cell Mol. Biol.* **2**: 151.
Wooding, F. B. P., and O'Donnell, J. M. (1971). *J. Ultrastruct. Res.* **55**: 71.
Yanagimachi, R., and Noda, Y. D. (1970a). *Amer. J. Anat.* **128**: 367.
Yanagimachi, R., and Noda, Y. D. (1970b). *J. Ultrastruct. Res.* **31**: 465.
Yanagimachi, R., and Noda, Y. D. (1970c). *Amer. J. Anat.* **128**: 429.
Yanagimachi, R., and Teichman, R. J. (1972). *Biol. Reprod.* **6**: 87.
Zahler, W. L., and Cleland, W. W. (1968). *J. Biol. Chem.* **243**: 718.
Zamboni, L., and Stefanini, M. (1968). *Fert. Steril.* **19**: 570.
Zamboni, L., and Stefanini, M. (1971). *Anat. Rec.* **169**: 155.
Zamboni, L., Zemjanis, R., and Stefanini, M. (1971). *Anat. Rec.* **169**: 129.
Zaneveld, L. J. D., Polakoski, K. L., and Williams, W. L. (1972). *Biol. Reprod.* **6**: 30.
Zeitler, E., and Bahr, G. F. (1962). *J. Appl. Phys.* **33**: 847.

INDUCTION OF CHROMOSOME CONDENSATION IN INTERPHASE CELLS

Potu N. Rao
and Robert T. Johnson

DEPARTMENT OF DEVELOPMENTAL THERAPEUTICS
THE UNIVERSITY OF TEXAS M.D. ANDERSON HOSPITAL AND TUMOR INSTITUTE
AT HOUSTON, HOUSTON, TEXAS
AND DEPARTMENT OF ZOOLOGY, UNIVERSITY OF CAMBRIDGE, CAMBRIDGE, ENGLAND

I. Introduction

In plant and animal cells, chromosomes can be visualized as discrete cytological entities only for a brief time during the life cycle, i.e., either during mitosis or meiosis. Throughout the entire intermitotic period, which varies from 10 to 40 hours for animal cells in culture, the chromosomes remain diffuse and indistinguishable in the nucleus. The amount of DNA present in a mammalian nucleus 6–8 μ in diameter just before mitosis, if placed tandem would reach a length of 4.5 meters (Stubblefield, 1973). In eukaryotic cells this great length of genetic message is distributed among a number of chromosomes. According to Stubblefield's (1973) calculations, the largest chromosome (12 μ long) in a Chinese hamster cell would contain about 45 cm of DNA. This DNA molecule, complexed with histone and nonhistone proteins, undergoes several orders of coiling, giving rise to chromatin fibers about 230 Å in diameter; these are commonly observed by electron microscopy in interphase nuclei or isolated metaphase chromosomes (Ris, 1956; Gay, 1956; Abuelo and Moore, 1969; DuPraw, 1968). Such 230 Å fibers are tightly packed into chromosomes at the time of cell division, a process that facilitates the equal distribution of genetic material between the two daughter cells.

It is almost a century since the chromosome cycle in plant and animal cells was first described (Flemming, 1880). Yet, key questions still remain unanswered: (1) How is the presumably ordered yet dispersed interphase chromatin organized into the dense chromosome bodies? (2) What is the nature of the intracellular signals that trigger dissolution of the nuclear membrane and condensation of chromosomes? Since chromosome condensation is invariably associated with mitosis, an understanding of this process might give us some insight into the regulation of cell division in normal and malignant cells.

In this review, we will discuss recent attempts to promote the condensation of chromosomes experimentally, together with some of the factors that may be involved in this process. The phenomenon of premature chromosome condensation (PCC), the fate and consequences of PCC, and the role of polyamines, cations, and proteins in the induction of PCC will be reviewed.

II. The Chromosome Condensation Cycle in Eukaryotic Cells

A. Normal Sequence of the Mitotic Cycle

In actively growing cells, the chromosomes go through a cycle of replication, condensation, anaphase separation, decondensation and subsequent reassembly into daughter nuclei. Yet there are no visible changes at the light microscope level, either in the cytoplasm or the nucleus, to indicate the status of a given cell with regard to the oncoming condensation of chromosomes or the initiation of mitosis. Increase in cell mass to a critical level is considered to be essential for cell division, but growth alone does not necessarily lead to division (Hertwig, 1908; Hartmann, 1928; Adolph, 1929; Scherbaum and Zeuthen, 1954; Prescott, 1955, 1956; Mazia, 1956; Swann, 1957; Mitchison, 1971). Owing to lack of any gross changes in the internal morphology of the cell during the intermitotic period, earlier cytologists referred to it as the "resting stage." Howard and Pelc (1953) using radioactive tracers and autoradiography, and Swift (1950) using Feulgen densitometry, demonstrated for the first time, the existence of a distinct period of DNA synthesis during interphase. On the basis of their data, the intermitotic period was divided into three phases, i.e., G_1, S, and G_2. The S phase represents the period of DNA synthesis, while G_1 and G_2 indicate the periods before and after DNA synthesis (Fig. 1).

The majority of eukaryotic cells conform to this general pattern of the mitotic cycle. In contrast to replicative DNA synthesis, which is confined to the S period, RNA and proteins are synthesized throughout the mitotic cycle; however, their rates of synthesis are lowest during mitosis, when the chromosomes are in a condensed state (Robbins and Scharff, 1966). The histones, which form one of the major constituents

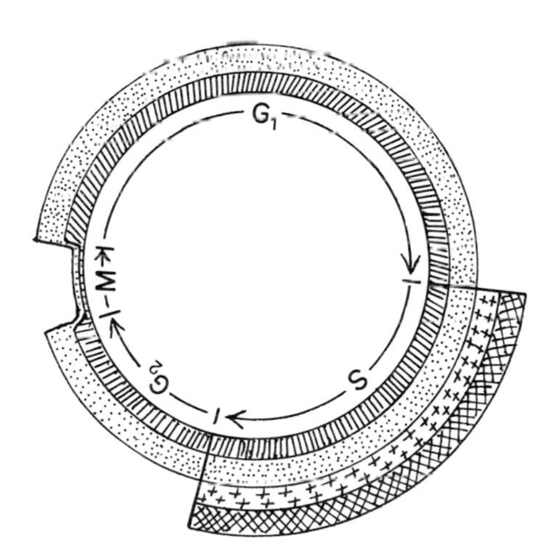

FIG. 1. Diagram of the mammalian cell cycle. G_1 = Pre-DNA synthetic period; S = DNA synthetic period; G_2 = Post-DNA synthetic period; M = mitosis. ▥, RNA synthesis; ⬚, protein synthesis; ⊞, DNA synthesis; ▩, histone synthesis.

of eukaryotic chromosomes, are synthesized in the cytoplasm at the same time as DNA synthesis is going on in the nucleus (Spalding *et al.*, 1966; Robbins and Borun, 1967; Borun *et al.*, 1967). Other chromosomal proteins appear to be synthesized during mitosis itself (Stein and Baserga, 1970).

Despite the stepwise or gradual synthesis of macromolecular components for chromatin during the cell cycle, little is known about the status of the chromosomes throughout interphase. The physical conformations of both interphase and mitotic chromosomes must be the net product of stable interactions between these macromolecular components and the ionic milieu of the nucleus. However, so far very little is known about the ionic environment of the nucleus throughout interphase and how this might influence the conformations of the chromatin fibers. Nevertheless, it is clear from studies with isolated histone-DNA fractions that very small changes in the concentrations of monovalent cations can change the binding relationship between certain histones and DNA (Ohlenbusch *et al.*, 1967). At mitosis, with the disruption of the nuclear envelope, marked changes must occur in the ionic environment around the chromosomes; perhaps as a result, rapid chromosome coiling is promoted and held stable for a time sufficient to permit the segregation of the genetic material equally into the daughter cells. There is some evidence that ferric ions (Fe^{3+}) may be involved in the formation or maintenance of the mitotic chromosomes (Robbins and Pederson, 1970), and it is also known that mitotic chromosomes contain considerable amounts of bound Ca^{2+} and Mg^{2+} ions (Steffensen, 1961; Cantor and Hearst, 1970). However, further studies are needed to assess the pattern of changes in the binding between chromatin and nuclear cations during the cell cycle.

Recent studies indicate that subtle changes take place in the organization of chromatin in the interphase nucleus. Pederson and Robbins (1972) have shown that the rate of binding of actinomycin D fluctuates in a cyclical pattern which bears some correspondence to the cell cycle (Fig. 2). In synchronized HeLa cells, the rate of actinomycin binding increases during G_1, reaches a peak during early S phase, and then slopes down to its lowest point at the time of mitosis (Pederson, 1972). Actinomycin binding to or intercalation into DNA may be indicative of the degree of chromatin and chromosome condensation. In any case, Fig. 2 indicates that the conformational changes taking place in the chromosomes during G_1 and early S phase are different from those occurring during the later part of the cell cycle.

Using pancreatic DNase I to probe the structure of chromatin isolated from synchronized HeLa cells, Pederson (1972) also demonstrated that

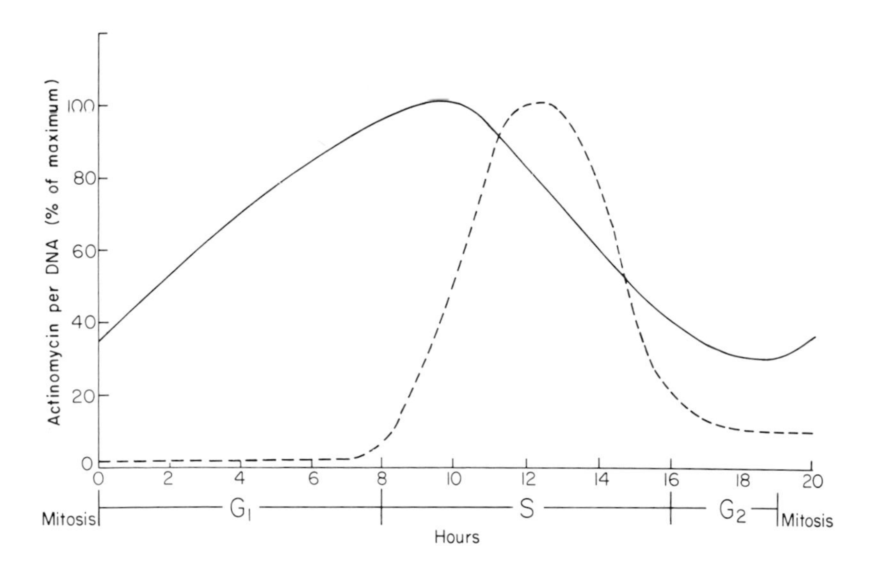

FIG. 2. Actinomycin binding in synchronized HeLa cells. The cyclic pattern shown was obtained regardless of whether ^{3}H-labeled actinomycin binding was measured in living cells, ethanol-fixed cells, or isolated nuclei *in vitro*. ^{3}H-labeled actinomycin binding (——) and DNA synthesis (---) are both shown as percentage of maximum. [From Pederson (1972). *Proc. Natl. Acad. Sci. U.S.* **69**, 2224, with permission.]

the sensitivity of DNA to DNase attack decreases as the cell progresses toward mitosis. This suggests two possibilities: (1) during G_1 there may be regions of DNA which are "naked", i.e., are not complexed with protein (Clark and Felsenfeld, 1971); or (2) following the replication of DNA, histone and nonhistone proteins (and perhaps other factors) accumulating during the S and G_2 phases interact with chromatin, causing structural changes in the chromatin which protect it from DNase attack. Additional evidence that conformational changes occur in chromatin during the chromosome cycle is provided by histochemical techniques. For example, Alvarez and Valladares (1972), using safranine staining in synchronized mammalian cells, have elegantly traced various changes occurring in the dye-binding capacity of interphase nuclei in relation to different phases of the cell cycle (Fig. 3). In addition, Nagl (1970) was able to correlate morphological changes in nuclei of *Allium carinatum* with the phase of DNA synthesis.

Taken together, these studies clearly indicate that during the cell cycle the chromosomes pass through a continuum of physical states whose two extremes, mitotic and S phase chromatin, reflect the greatest and least degrees of condensation, respectively. Admittedly these ideas are not new; Mazia (1963) and others have proposed chromosome condensa-

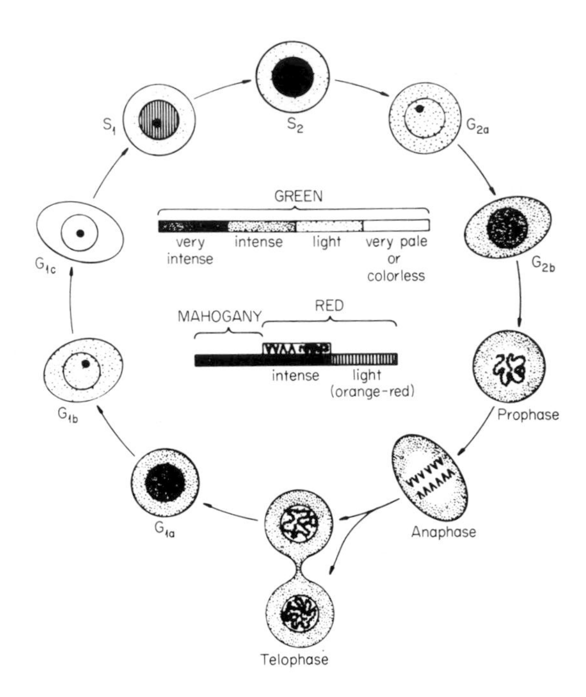

FIG. 3. Differential staining during stages of the cell cycle. [From Alvarez and Valladares (1972). *Nature New Biol.* **238**, 279.]

tion-decondensation cycles with respect to the cell division cycle. However, accumulating evidence fits well with the available models.

Mitosis

Data presented in the preceding section suggest that, immediately after DNA replication, the process of chromosome condensation begins, even though it cannot be detected by simple light microscopy. This may be considered the first order of condensation (further evidence will be presented in Section VII to support this assumption). The second order of condensation, which is more rapid and dramatic, begins with the initiation of mitosis. In this section mitosis will be discussed primarily with respect to chromosome condensation. On the basis of light and electron microscopic studies, the various events of mitosis can be summarized as follows:

a. Prophase. The first obvious sign of the initiation of mitosis is the condensation of chromatin at the periphery of the nucleus (Fig. 4). The nuclear envelope invaginates to a certain extent, producing a nuclear

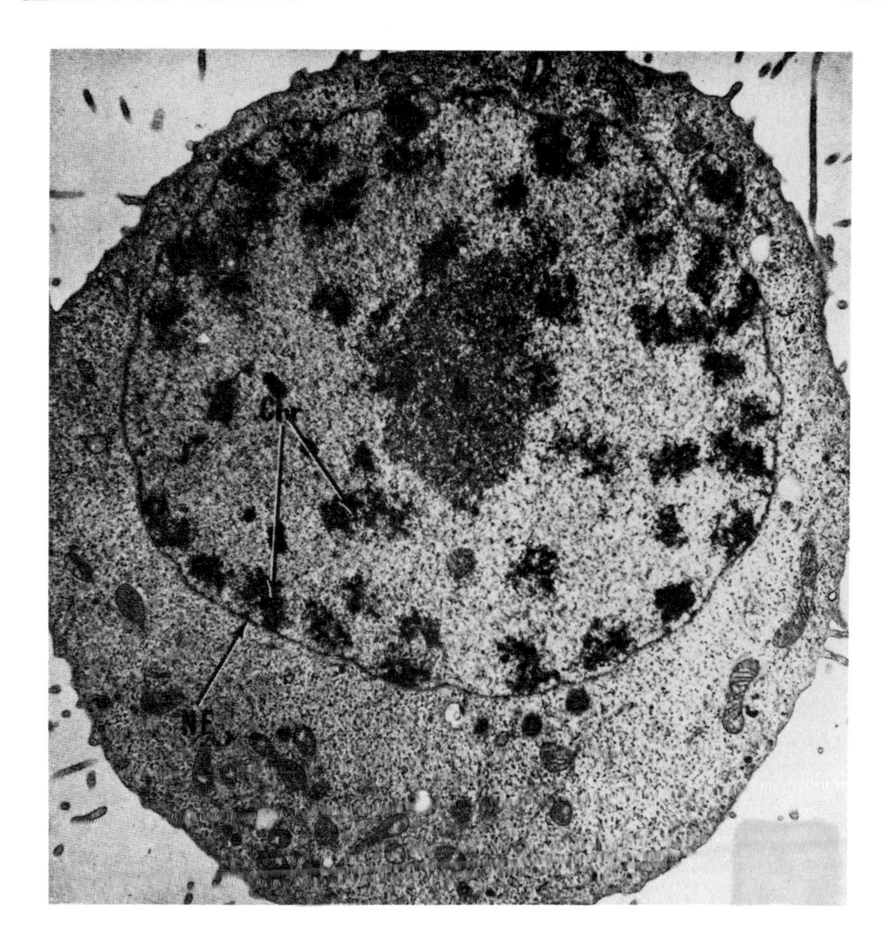

FIG. 4. Thin-section electron micrograph of a HeLa cell in prophase. Note the distribution of condensed chromatin and the nucleolus within the nucleus. ×7000. [From Robbins *et al.*, (1970). *J. Cell Biol.* **44**, 400.]

"hof" which contains the centriole pair (Robbins and Gonatas, 1964). Separation of the centrioles leads to the establishment of poles during prophase. The chromatin network becomes more distinct, until gradually long, slender prophase chromosomes become recognizable. Spindle microtubules originating from the centrioles extend to the intranuclear chromosomes, even though no distinct kinetochore regions are detectable. Disintegration of the nuclear membranes and disappearance of the nucleolus mark the end of prophase.

Evidently just at this time there is a rapid association of RNA with the prophase chromosomes (Kaufmann *et al.*, 1948; Jacobson and Webb,

1952; Boss, 1954). Because of the apparent coincidence of nucleolar breakdown and the binding of RNA to the chromosomes, Brown (1954) thought that the chromosomal RNA might be derived from the nucleolus. However, other studies suggest that some of the chromosomal RNA could be of cytoplasmic origin (Jacobson and Webb, 1952; Boss, 1955; Love, 1957; Huberman and Attardi, 1966). There is some controversy over whether RNA is an integral part of the chromosomes (Bonner *et al.*, 1968), or whether it is only associated with them in a superficial manner (Huberman and Attardi, 1966). The completion of prophase is also associated with a massive loss of nonchromosomal proteins from the nucleus (Richards and Bajer, 1961; Prescott and Bender, 1963).

b. Metaphase. By the end of prophase, the chromosomes are fairly well condensed and become engaged to the mitotic spindle. However, throughout metaphase and even up to early anaphase, chromosome condensation continues (Bajer, 1959). Time lapse studies of mitosis in *Haemanthus* endosperm reveal that the length of a given chromosome during anaphase is only about half of its prophase length. A fully condensed metaphase chromosome appears to be a bundle of folded chromatin fibers (Fig. 5; factors that may be responsible for such folding will be discussed in Section VI). During metaphase the compact and highly condensed chromosomes line up on the equatorial plane of the mitotic spindle. The two sister chromatids of each metaphase chromosome are held together only at the undivided centromere region (Fig. 5).

c. Anaphase. The beginning of anaphase is marked by a division of the centromeres and a gradual but speedy movement of the daughter chromosomes toward the two opposite poles. As soon as the daughter chromosomes reach their respective poles, the decondensation process may commence.

d. Telophase. Chromosomal division is soon followed by cytokinesis, or division of the cytoplasm. Then the sequence of events that occurred during prophase is exactly reversed during telophase. The chromosomes undergo decondensation to such an extent that they are no longer distinguishable cytologically. The nuclear envelope is reassembled, and the nucleolus reappears. RNA and protein liberated from the prophase nucleus migrate back into the reconstituted daughter nuclei (Prescott and Goldstein, 1968; Neyfakh *et al.*, 1971). It seems, however, that decondensation of the chromosomes is not complete until late G_1 or early S phase, as measured by the rate of actinomycin binding to chromatin (Fig. 2).

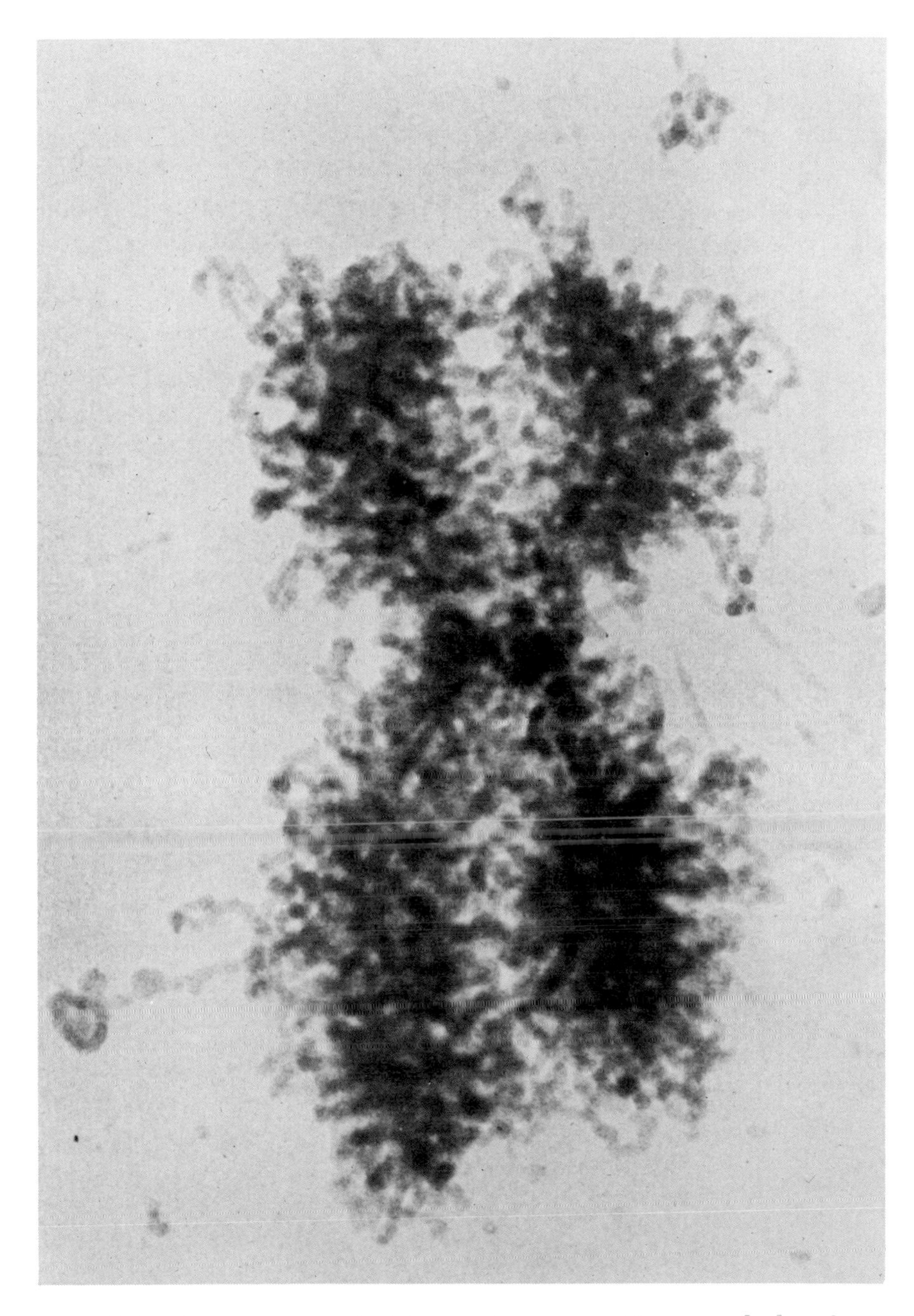

FIG. 5. An electron micrograph of human chromosome 16 in a standard configuration. ×46,300. [From DuPraw (1970). "DNA and Chromosomes." Holt, Rinehart and Winston, Inc. New York.]

B. Exceptions to the Normal Chromosome Cycle

In general, the chromosomes of eukaryotic cells exist in a diffuse state during interphase. However, there are a few exceptions to this rule, such as the giant polytene chromosomes of certain Diptera. Cells that contain these polytene chromosomes are in interphase, not in mitosis. The giant chromosomes are unique in that they are made up of a large number of unit chromatids lying in parallel arrays (DuPraw and Rae, 1966). Each chromosome is reduplicated about 1000 times, and hence they are basically polyploid in nature. Various banded regions contain DNA more closely packed than the interband regions (DuPraw, 1968). However, visibility of these chromosomes by light microscopy at interphase is due to the massive reduplication of basic chromosomal units, rather than to the type of chromosome condensation seen during the initiation of mitosis.

A second example is found in certain flagellates, including Holomastigotoides, in which the chromosomes remain coiled and visible throughout interphase (Cleveland, 1949). Soon after cell division, these chromosomes undergo decondensation, replication, and then condensation; thereafter, they remain in a condensed state for a long time before the next cell division. In these species chromosome condensation is not followed immediately by organization of a spindle and separation of sister chromatids. This suggests that chromosome condensation and spindle formation, which usually are coordinated and occur concurrently, may be uncoupled in rare instances, e.g., in the case of Holomastigotoides. Another example is the induction of premature chromosome condensation (PCC) by fusing mitotic with interphase cells; here again, chromosome condensation does not appear to be associated with spindle formation (Johnson and Rao, 1970).

Still another form of interphase chromosome condensation, which is seen in virtually all somatic cells, is known as heterochromatinization (Lima-de-Faria, 1959; Brown, 1966). When a region of the genome becomes heterochromatic, it generally loses most if not all of its template activity, and it tends to replicate late during the S phase. One of the best known examples is the differential inactivation of one of the two mammalian X chromosomes in female somatic cells (Lyon, 1962, 1972). However, in some species, particularly rodents, only part of the X chromosome is inactivated in this manner (Ohno, 1969). The mechanism by which certain chromosomes or regions of chromosomes can be specifically inactivated and condensed, while the rest of the genome is active and euchromatic, remains unknown. It has been suggested that a relationship exists between nuclear volume and heterochromatinization

(Mittwoch, 1967); also, it is clear that the heterochromatic areas are often associated with the nuclear envelope. The observation that such heterochromatic regions of the genome become euchromatic in very early embryos strongly suggests that regulation of this mode of condensation differs essentially from the general condensation that occurs at prophase.

III. Effects of Chemical Agents in Chromatin Condensation

In this section we will review various attempts to mimic the nuclear and cytoplasmic changes seen during mitosis by various treatments, such as hypertonicity or the administration of basic compounds, e.g., the aliphatic polyamines. Table I summarizes some results of such treatments, using both whole cells and isolated nuclei.

From a review of the earlier literature (Anderson, 1956), together with the information in Table I, we may conclude that indeed many degrees of chromatin aggregation, as well as other mitosis-associated phenomena, can be achieved by effectively increasing the cation concentration (Fig. 6). None of these data, however, clearly demonstrate that a true prophase condensation has been induced, i.e., one leading to dissolution of the nuclear membrane and appearance of recognizable chromosomes. Indeed, evidence provided by Philpot and Stanier (1957) suggests that the nature of chromatin condensation in an actual prophase nucleus differs in an undetermined manner from chemically induced condensation.

Our attempts to induce chromosome condensation in interphase HeLa cells by increasing the ionic strength of the medium resulted in nonspecific and irreversible condensation of chromatin, but did not produce chromosomes. Incubation of a random population of HeLa cells at 37°C in a buffer containing 10^{-2} *M* Tris·HCl; 10^{-2} *M* Mg^{2+} (as $MgCl_2 \cdot 6H_2O$); 0.2 *M* KCl, and 0.006 *M* β-mercaptoethanol (with pH adjusted to 7.4 by adding Tris base) caused a certain degree of condensation. This chromatin condensation became even more marked when synchronized G_1 phase cells were incubated for 60 minutes with 1000 HAU of UV-inactivated Sendai virus in a total volume of 1 ml of the above buffer (Fig. 7). During the initial stages, a network of chromatin formed, bringing the nucleoli into prominence (Fig. 7A). Later the nuclear envelope and nucleoli disappeared, leaving an aggregate of condensed chromatin (Fig. 7B). Resuspending this material in isotonic medium did not lead to decondensation or to the formation either of a nucleus or of individual chromosomes. In a similar experiment, substituting

TABLE I
EFFECT OF CHEMICAL AGENTS ON CHROMATIN CONDENSATION

Whole cell	Isolated nucleus	Agent used	Chromatin condensation	Recovery	Other Pseudomitotic effects	References
HeLa		1.6–2.8× isotonic NaCl, $MgCl_2$, KCl	Prophaselike condensation around nucleolus and nuclear envelope. Wider variation of granulation than in normal prophase	Complete reversal in isotonic medium	Dispersion of polyribosomes. Dissolution of nuclear envelope at 2.8× isotonicity.	Robbins *et al.* (1970)
	HeLa	2.4× isotonic NaCl	None		None	Robbins *et al.* (1970)
	HeLa	2.8× isotonic NaCl, KCl, $MgCl_2$, $CaCl_2$	None; prophaselike chromatin followed by nuclear dissolution and extrusion of DNA. Indiscriminate precipitation of chromatin			Robbins *et al.* (1970)
	(Rat liver parenchyma)	0.01–0.35 *M* *n*-propylamine	Nuclear swelling at low concentrations			Anderson and Norris (1960)
	(Rat liver parenchyma)	Putrescine, cadaverine, agmatine, lysine, or arginine 0.001 *M*-0.5 *M*	Nuclear shrinkage, granulation and nuclear condensation in 0.025 *M* solution	Complete reversal by distilled H_2O or isotonic medium		Anderson and Norris (1960)
		Spermidine, 0.001 *M* or Spermine, 0.001 *M*	Marked condensation to produce bodies whose numbers are approx. the same as the chromosomes of the species	Only slowly reversible in PO_4-containing solution.		Anderson and Norris (1960)

	Rat liver parenchyma	NaCl 0.05–0.5 M	Fine granulation	Reversible by placing nuclei in isotonic medium	Anderson and Wilbur (1952)
		KCl 0.05–0.5 M	Fine granulation		
		$CaCl_2$ 0.25 M	Marked granulation		
		$MgCl_2$ 0.25 M	Granulation less than with Ca^{2+}		
		Decrease in pH 6.1 → 4.64	Fine granulation at pH 5.5		
Grasshopper neuroblasts		Agmatine 0.005 M	Acceleration of cells in prometa-metaphase, early middle telophase, and also through G_2		St. Amand *et al.* (1960)
Grasshopper neuroblasts		$CaCl_2$ 0-concentration	Prolongation prometa-metaphase		St. Amand *et al.* (1960)
		0.0017 M (control)	Normal passage through mitosis		
		0.017 M	Prolongation of prometa-metaphase		
Rat thymocytes (normal and irradiated)		$CaCl_2$ 0.12–2 mM; agmatine 2.5 mM	Increased flow of cells into mitosis		Whitfield and Youdale (1966)
	Rat liver parenchyma	Increase in ionic strength from 0.03 → 1.03; decrease in pH 7.0 → 5.1; $MgCl_2$ 0.008 M, protamine 0.02%, histone 0.05%	All produce a granularity which differs from true prophase condensation in the fineness of the effect	Reversible by triphosphate	Philpot and Stanier (1957)

(*Continued*)

TABLE I (*Continued*)

Whole cell	Isolated nucleus	Agent used	Chromatin condensation	Recovery	Other Pseudomitotic effects	References
Chinese hamster (DON)		Arginine-rich histone (50–2000 μg/ml)	Slender chromatin networks		Nuclear membrane dissolution	Matsui *et al.* (1971)
		Other histone fraction	None			Matsui *et al.* (1971)
	Chinese hamster (DON line)	Arginine-rich histone (50–2000 μg/ml)	None			Matsui *et al.* (1971)
Krebs ascites tumor (mouse)		Protamine, calf thymus histone (50–300 μg/ml)	Granularity in the nucleus	Nil		Becker and Green (1960)
Sea urchin eggs		Formamide (1 *M*)	Chromosomes lose form and collapse into spheres	Upon removal from formamide		Nevo *et al.* (1970)
	Formalin-fixed nuclei of *Vicia faba* root tips	Trypsin (1 mg/ml)	Duplex chromosomes appear from G_2 nuclei, but not from G_1			Wolff (1969)

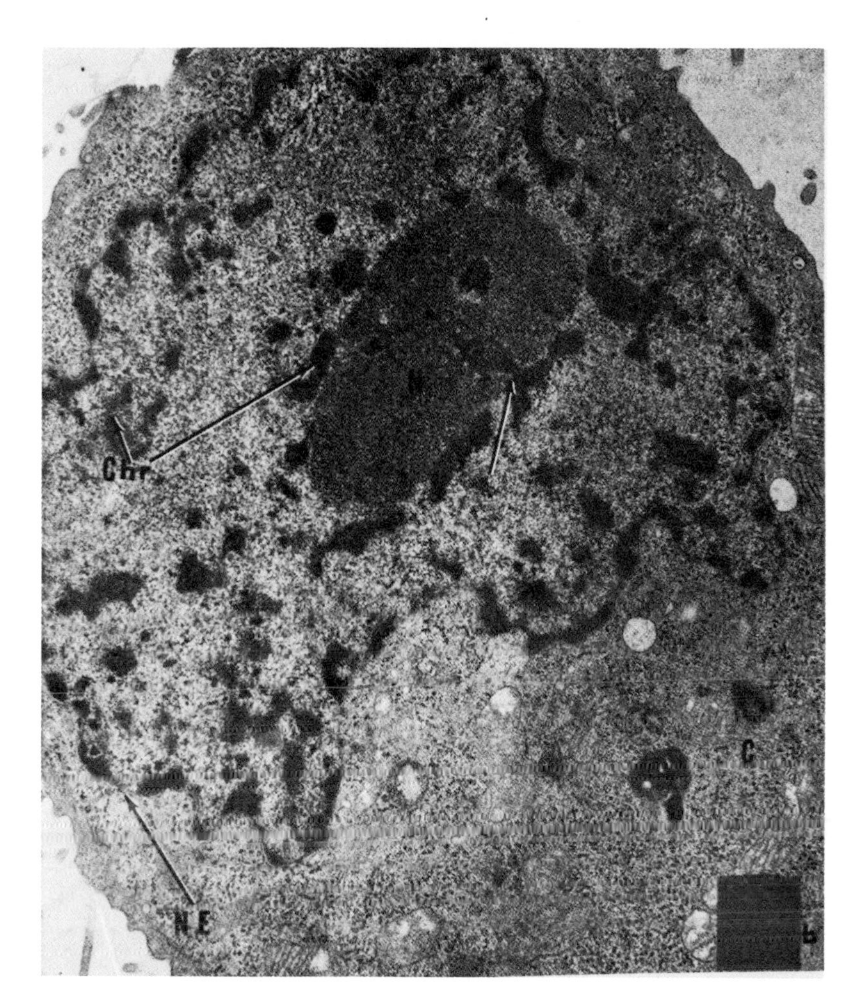

FIG. 6. Interphase cell treated for 2 minutes with 1.6 × isotonic medium. Chromatin (Chr) is condensed preferentially along the inner aspect of the nuclear envelope (NE) and around the nucleolus. The nuclear envelope is ruffled. The nucleolus shows diffuse granularity and the absence of fibrous component. Polyribosomes are intact. At the arrow, condensed intranucleolar chromatin is continuous with chromatin in the nucleoplasm proper. ×10,000. [From Robbins *et al.,* (1970). *J. Cell Biol.* 49, 400.]

Hank's salt solution for the hypertonic buffer, no such condensation was observed.

Condensation of interphase chromatin to form true chromosomes has long been thought to be due to an interaction between various basic molecules and acidic DNA (Heilbrunn, 1952; Anderson, 1956). However,

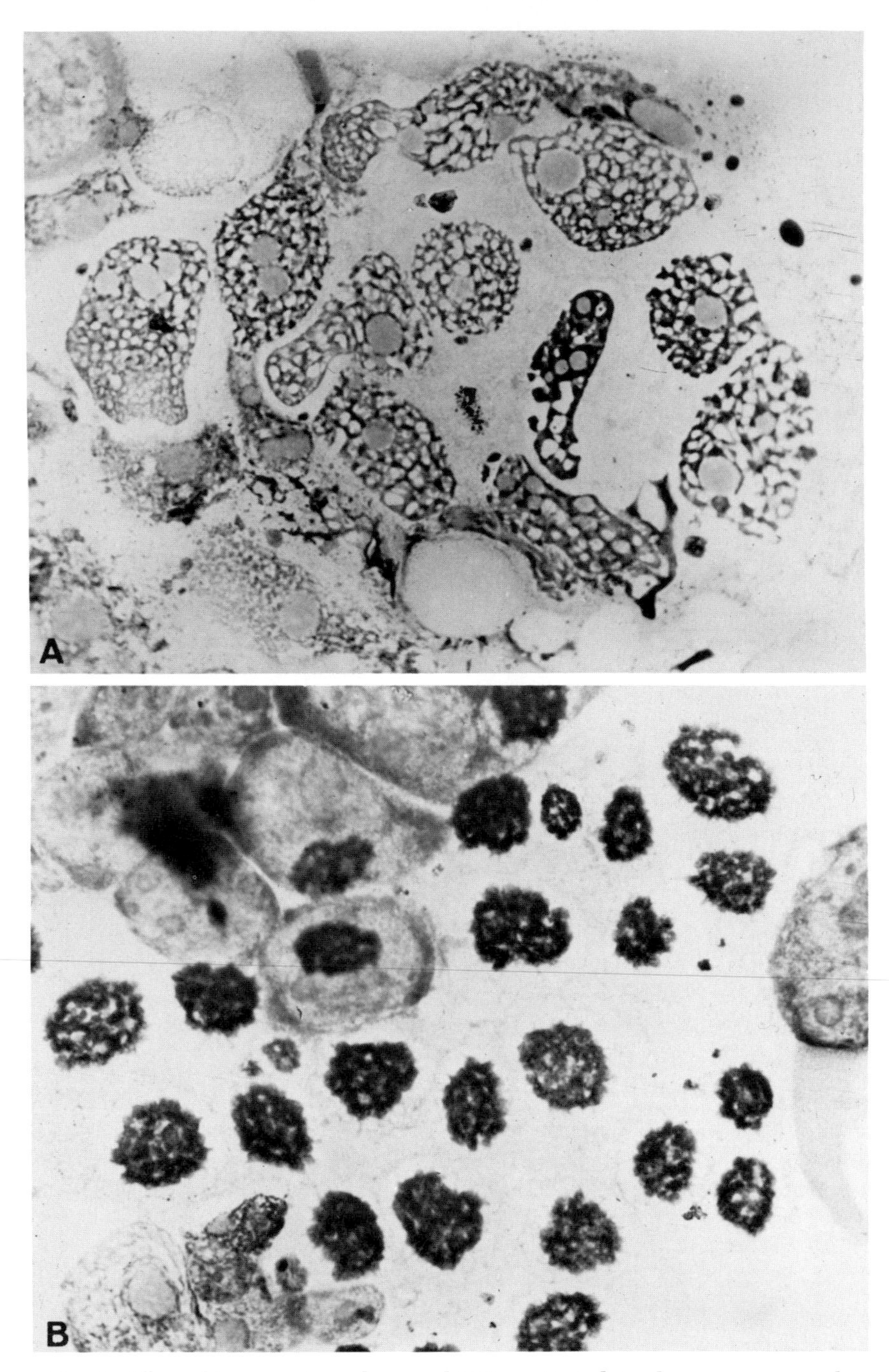

FIG. 7. Effect of hypertonic medium and UV-inactivated Sendai virus on a synchronized population of HeLa in G_1 phase. At (A) 15 minutes, (B) 60 minutes after the addition of virus and medium.

the mechanisms controlling shifts in anion–cation balance in cells, or in compartments within cells, are by no means clear. Induction of chromatin condensation by elevated concentrations of monovalent and divalent cations, or by multicationic polyamines, results from a shift in the anion–cation balance (Table I); however, this treatment must be given suddenly, since a gradual increase in concentration has little effect (Philpot and Stanier, 1957).

The role of the larger aliphatic polyamines, spermidine and spermine, in the condensation of chromatin remains unresolved. Although these compounds produce the greatest degree of nuclear condensation, it is generally irreversible (Davidson and Anderson, 1960). This has become apparent both at the light and electron microscopic levels of investigation. For example, Chevaillier (1969), working with the histone-free chromatin of nuclei from a crab, found that the addition of both spermidine and spermine produced thickening and cross-binding of chromatin fibers. Molecular models for the binding of spermidine and spermine to DNA have been proposed (Liquori *et al.*, 1967; Glaser and Gabbay, 1968; Suwalsky *et al.*, 1969); on the other hand, no clear evidence exists that these compounds bind in a similar manner *in vivo*, nor that they are responsible for the ordered condensation of chromatin.

The role of histones in chromatin condensation is also unresolved. It has been shown by many workers (Whitfield and Youdale, 1966; Scaife and Brohee, 1969) that X-irradiation of tissue culture cells causes delayed entry into mitosis. One effect of irradiation is the disaggregation of chromatin in the interphase nucleus (Whitfield *et al.*, 1962), and this may be causally linked to a release of histones from chromatin (Hagen, 1960; Gurley *et al.*, 1970). Since the radiation-induced delay can be overcome by addition of Ca^{2+} or agmatine, it has been suggested that these cations prevent uncoiling and dissolution of the chromatin aggregates in the nuclei, and thus promote mechanisms by which cells enter into prophase condensation (Whitfield and Youdale, 1966). Addition of histones or protamines either to isolated nuclei (Philpot and Stanier, 1957) or to intact cells (Becker and Green, 1960; Matsui *et al.*, 1971) also causes chromatin condensation, but in neither case are true chromosomes obtained.

In general, chemical analysis of interphase and metaphase chromatin has not revealed any substantial differences either in the quantity or quality of their histone composition (Comings, 1967; Maio and Schildkraut, 1967; Hancock, 1969). However, Sadgopal and Bonner (1970) did find that metaphase chromosomes are deficient in lysine-rich histones. Recently it has been reported by Lake *et al.* (1972) that the F1 histone fraction in Chinese hamster cells is phosphorylated exclusively during metaphase; how this might influence chromosome condensation is at present unresolved.

The above data, taken together, indicate that a wide range of cationic molecules are capable of inducing a prophaselike transition in interphase nuclei. However, no treatment has yet resulted in the production of true chromosomes. It must be concluded that other factors are involved in the normal condensation phenomenon (Section VI,B).

IV. Cell Fusion and Chromosome Condensation

A. Premature Chromosome Condensation

Initially we set out to study the pattern of mitotic accumulation among binucleate HeLa cells, which were formed by the fusion of S phase cells (prelabeled with thymidine-^{3}H) with unlabeled G_2 cells (Rao and Johnson, 1970). Unexpectedly, we observed that, in some of the heterophasic binucleate cells, a normal set of unlabeled chromosomes was associated with a "pulverized" set of labeled chromosomes (Johnson and Rao, 1970). Similar examples of chromosome pulverization had been reported earlier (Stubblefield, 1964; Sandberg *et al.*, 1966; Kato and Sandberg, 1967, 1968a,b,c; Takagi *et al.*, 1969). Our observation of pulverized chromosomes in S/G_2 fused cells suggested that advanced entry of the G_2 component into mitosis might have caused abnormal condensation within the S phase nucleus.

This possibility led us to perform another experiment, in which mitotic HeLa cells blocked with Colcemid were fused with synchronized populations of either G_1, S, or G_2 cells (Johnson and Rao, 1970). Fusion between mitotic and interphase cells resulted in rapid chromosome condensation, with dissolution of the nuclear envelope within 30 minutes after treatment with inactivated Sendai virus. This precipitous induction of chromosome formation from an interphase nucleus has been termed "premature chromosome condensation" (PCC) (Johnson and Rao, 1970). The morphology of the prematurely condensed chromosomes varies according to the stage of the interphase cells in the cell cycle at the time of fusion. Thus, the prematurely condensed chromosomes of G_1 phase are very long single chromatids, those of G_2 are elongated and slender double chromatids, and those of the S phase are characterized by their fragmented appearance. Figure 8 compares typical G_1, S, and G_2 prematurely condensed chromosomes with normal metaphase chromosomes of the Muntjak cells.

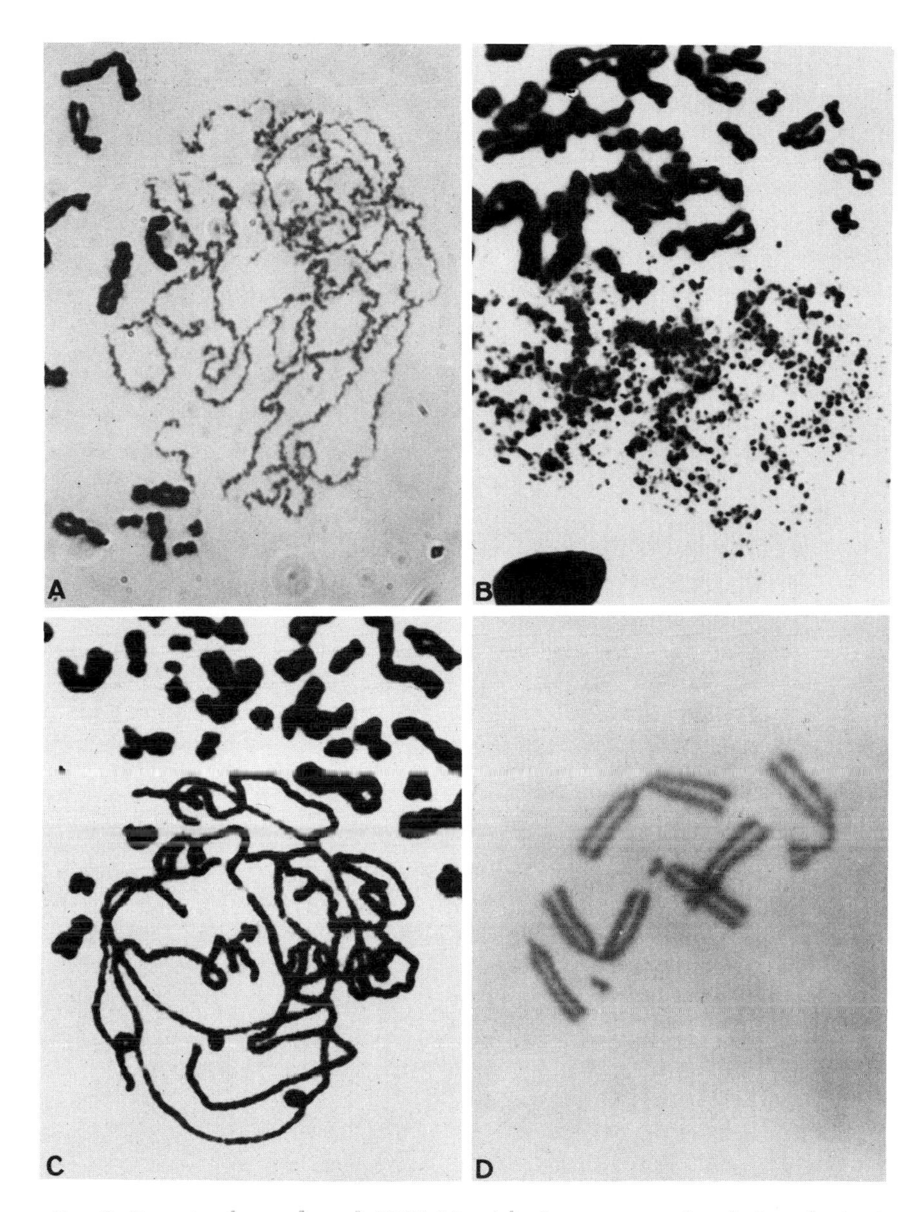

FIG. 8. Prematurely condensed (PC) Muntjak chromosomes after fusion of mitotic Chinese hamster cells with a random population of Muntjak cells. (A) G_1 PC chromosomes; (B) S phase pulverized PC chromosomes; (C) G_2 PC chromosomes; and (D) metaphase chromosomes. (From Hittelman and Rao, unpublished data.)

Changes in the interphase nucleus occur almost immediately after heterokaryons are formed. To a very great extent these changes mimic the normal prophase condensation patterns (Kato and Sandberg, 1968a; Johnson and Rao, 1970; Matsui *et al.*, 1972; Aula, 1970) although certain differences have been noted (Matsui *et al.*, 1972). Within 30 minutes after the completion of cell fusion, the interphase nucleus undergoes characteristic changes which are independent of the precise stage which it previously occupied. These earlier changes include chromatin condensation at the nuclear envelope (Fig. 9), which often brings the nucleoli into prominence (Fig. 10A, B). After these initial changes, condensation continues leading to dissolution of the nuclear membrane, and finally the disappearance of the nucleoli (Fig. 10C–F). The final result of PCC, of course, depends upon the phase of the interphase cell at the time of fusion.

Based on present information, we can make the following generalizations regarding the PCC phenomenon.

1. The process of condensation starts in the interphase nucleus immediately after fusion with a mitotic cell. In general, the incidence of PCC reaches a maximum within 1 hour after the addition of virus (Johnson and Rao, 1970; Kato and Sandberg, 1968a).

2. The speed of induction, and the extent to which it occurs in the fused cells, is closely related to the ratio of mitotic to interphase elements in the cell at the time of fusion (Johnson and Rao, 1970).

3. There are three basic types of prematurely condensed chromosomes, which correspond exactly with the three divisions of interphase in the cell cycle. The G_1 type consists of single chromatids, which are extremely long and usually present a coiled appearance (Fig. 8a). The G_2 type consists of double chromatids that remain close together, and are essentially normal replicated chromosomes; however, they are greatly extended and in general much longer than prometaphase elements. S-phase prematurely condensed chromosomes, over which much confusion has arisen (see Section IV,B), present a variety of appearances according to how far the particular S-phase nucleus has progressed in the course of DNA synthesis. For example, an S phase condensation may appear as a mass of large and small fragments (Fig. 8B), or as a mixture of thin, nonreplicated G_1 chromatids. Gaps are also seen, possibly representing the replicating regions of S phase. Finally, fully replicated, dense duplex G_2 regions are also seen, as shown in Fig. 11 (Johnson and Rao, 1970; Johnson *et al.*, 1970; Stenman and Saksela, 1971; Matsui *et al.*, 1972).

4. Apparently there are no species barriers for the induction of PCC. A mitotic cell from one species is capable of inducing PCC following

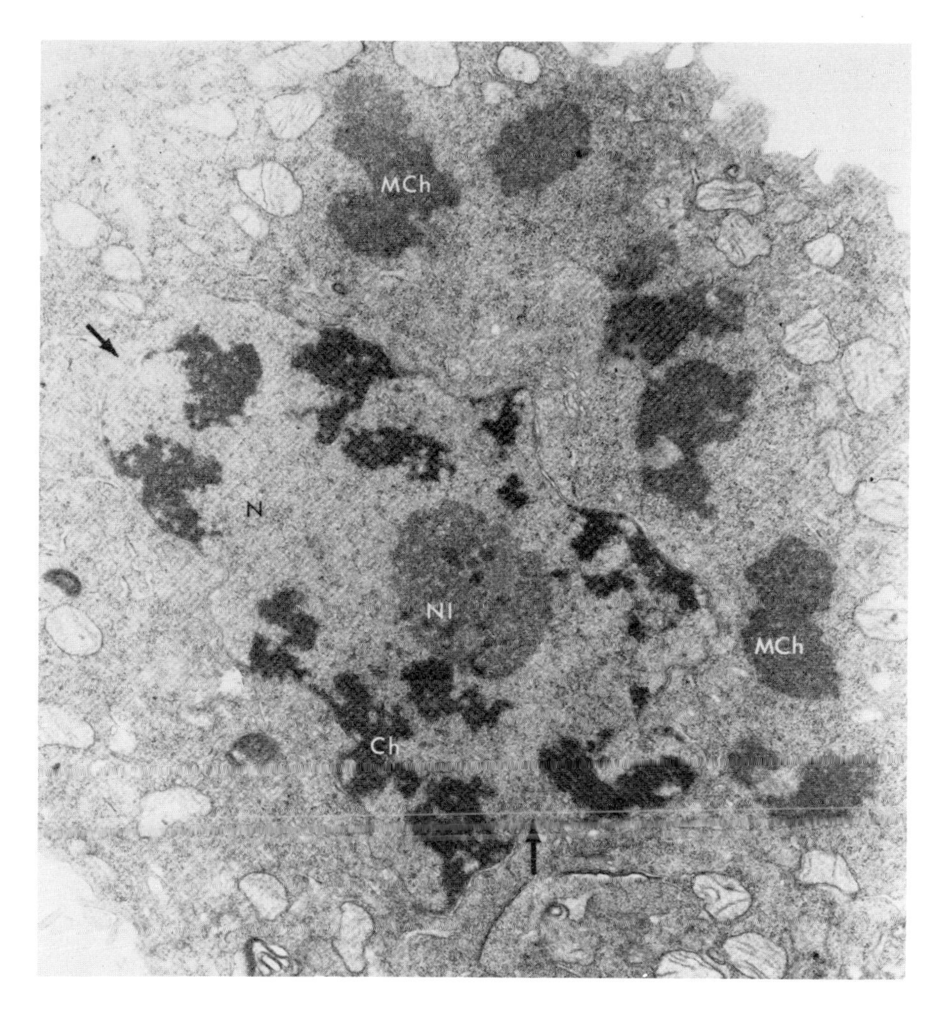

FIG. 9. Electron micrograph of a mitotic-interphase fused cell at 10 minutes after cell fusion. The chromatin (Ch) is distinctly condensed, some of it lying along the inner aspect of the nuclear membrane. Part of the nuclear envelope is disrupted (arrows). The nucleolus (Nl) is composed predominantly of granular elements. Metaphase chromosomes appear intact (MCh). In the cytoplasmic area, polysomes are now rare. N, nucleus. ×14,400. [From Matsui *et al.* (1972). *J. Cell Biol.* **54**, 120.]

fusion either with a highly differentiated cell (Johnson *et al.*, 1970) or with an undifferentiated cell of another species (Johnson *et al.*, 1970; Ikeuchi and Sandberg, 1970).

5. The number of G_1 or G_2 prematurely condensed chromosomes cor-

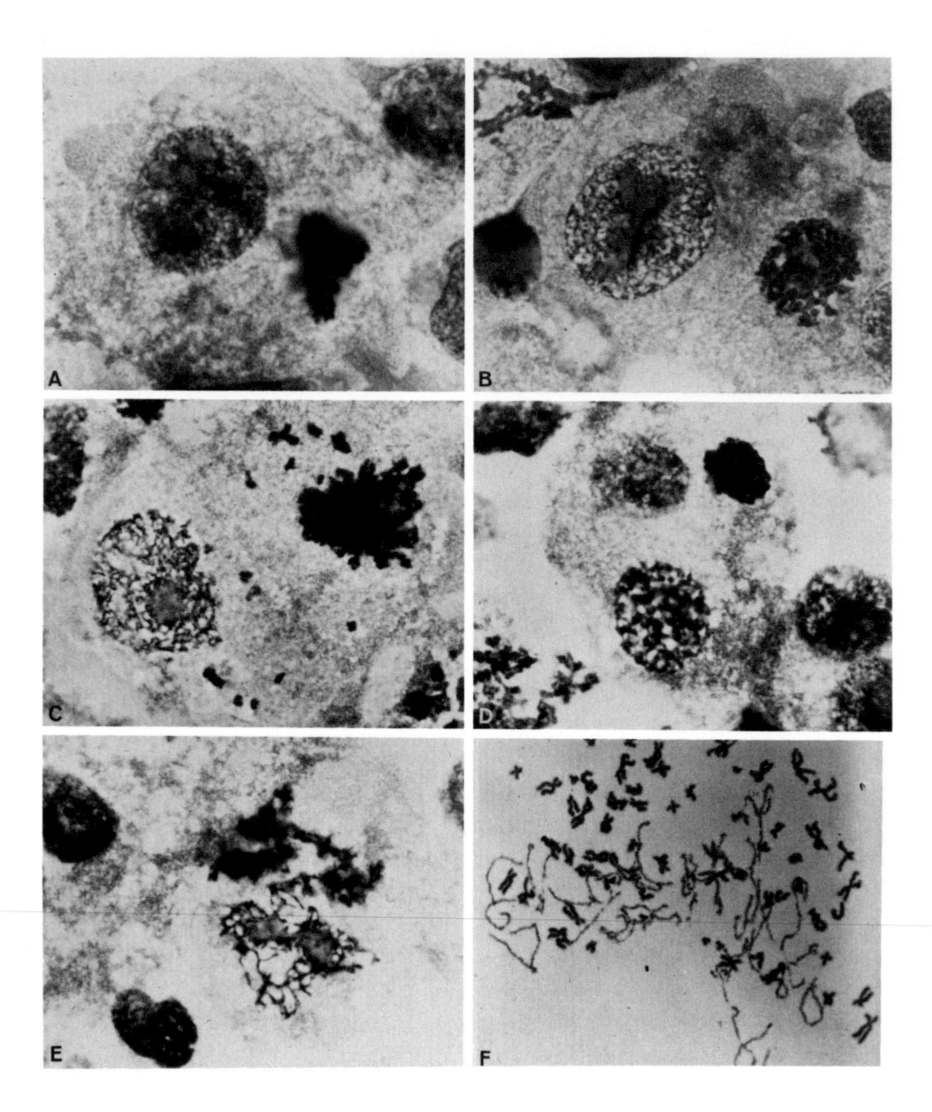

FIG. 10. Sequence of induction of premature chromosome condensation (PCC) in an interphase nucleus after fusion with a mitotic cell. (A) to (E) show various degrees of nuclear condensation. (F) shows PCC of the G_1 phase.

responds to the chromosome number of the species from which the interphase cells are obtained (Johnson *et al.*, 1970). It should be noted that an exactly similar PCC was described by Gurdon (1968), who injected adult and embryonic nuclei into *Xenopus* oocytes which were in meiosis. The interphase nuclei were rapidly induced to form chromo-

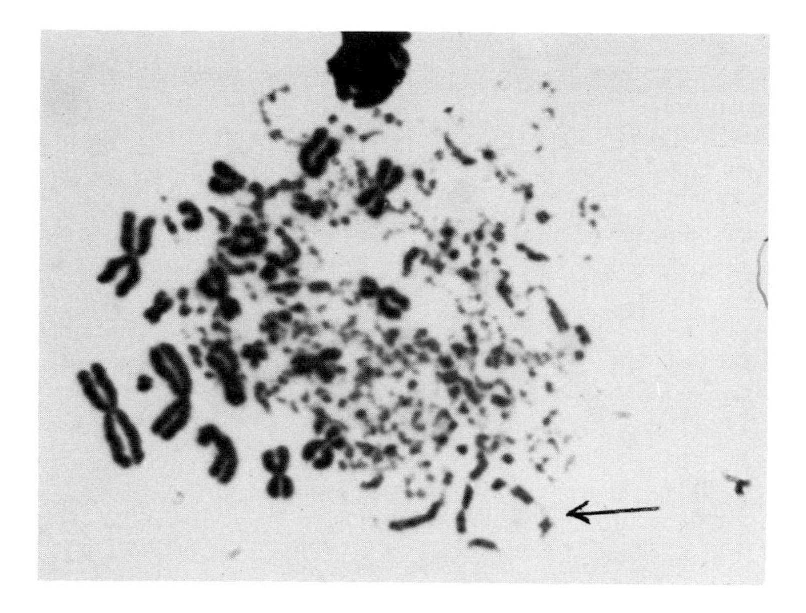

FIG. 11. Premature chromosome condensation of an S phase nucleus in a M/S fused cell. The arrow points to a chromosome in which some regions were replicated, while others were not. Replicated regions appear thick and heavily stained.

somes, suggesting that meiotic condensation has much in common with the process of mitotic condensation.

B. THE STRUCTURE OF PREMATURELY CONDENSED CHROMOSOMES

When chromosome preparations of mitotic-interphase fused cells are made within 15 minutes after the addition of virus, then many interphase nuclei are only partially condensed; this is reflected in the fineness and extreme length of both G_1 and G_2 elements. These early inductions resemble tangled arrays of fibers, and it is extremely difficult to identify by means of light microscopy which stage of interphase such nuclei occupied at the time of fusion.

The morphology of the S phase PCC ("pulverization" in the terminology of some workers) is again closely associated with the precise stage of the interphase nucleus within S phase at the time of fusion. For example, we have described a transitional morphology in S phase nuclei which had been held at the G_1-S boundary by excess thymidine (Johnson and Rao, 1970). Although some fragments were seen, the condensed nuclei contained many G_1 chromatids. This type of morphology has also been described by Stenman and Saksela (1971). The farther the interphase nucleus has traversed through S phase, the greater are the chances of discovering fully replicated regions of the genome which have a G_2 (i.e., double) appearance (Johnson and Rao, 1972). It should be stressed that the normal S phase nucleus, at any given moment, is a composite of S, G_2, and G_1 elements, which can be identified clearly in many preparations of S-phase PCC (Fig. 11). In addition, both the length of time that the induction is allowed to continue, and the ratio of mitotic to interphase components, also determine the degree of condensation.

Frequently the structure of S-phase PCC can be seen more clearly after quinacrine staining (Patil, Rao, and Johnson, unpublished; Johnson and Skaer, unpublished). In these cases, it is possible to see skeins of fine fluorescing strands, which appear to link up with larger fragments. The most important point to be ascertained is whether the replicating portion of the genome can be visualized, and whether or not the fragments seen under the light microscope are really isolated fragments. After staining for constitutive heterochromatin according to the method of Arrighi and Hsu (1971), many of the fragments in HeLa S-phase PCC exhibit C-bands. However, as compared with G_1 and G_2 PCC, in which C-band heterochromatin occurs as a few large blocks with specific locations (see below), there appears to be some fragmentation of the C-bands in S-phase PCC (Unakul *et al.*, unpublished data). This may be related to the replicating state of the chromosomes at this stage (Johnson and Rao, 1970).

When these points are taken into account, the terminology used to describe the various morphologies of prematurely condensed chromosomes may not be too complicated. The classification of Sandberg *et al.* (1970), which divides prematurely condensed chromosomes into eight groups, can be reconciled both with the findings of Johnson and Rao (1970) and those of Stenman and Saksela (1971), provided that cells at the boundary of G_1-S, the variable nature of S itself, and possibly cells in S-G_2 transition are recognized as such.

G_2 prematurely condensed chromosomes are essentially normal, though much elongated, chromosomes. Although the G_1 chromatids are probably also "normal," this needs further confirmation. Criteria of normality depend upon a number of features in the prematurely condensed chromosomes: (1) The number of condensed G_2 or G_1 elements should correspond to the expected number of metaphase chromosomes in that cell type; (2) constitutive heterochromatin should be located at similar regions in prematurely condensed chromosomes as in metaphase chromosomes; (3) C-banding patterns should be present in G_2 prematurely condensed chromosomes, and these should be identical in homologous chromosomes; (4) cross-banding patterns in G_2 prematurely condensed chromosomes should be similar to the patterns found in corresponding metaphase chromosomes.

Although none of these criteria has been exhaustively proved, there is sufficient evidence to predict that each criterion is satisfied in prematurely condensed chromosomes. In particular, the number of G_1 and G_2 condensed elements is the same as the metaphase chromosome counts for a number of cell types (Johnson *et al.*, 1970); this applies not only to proliferative tissue culture cells, but to differentiated cells as well. The C-banding, or constitutive heterochromatin patterns, in G_1 and G_2 HeLa prematurely condensed chromosomes also appear to correspond closely with metaphase C-bands (Fig. 12A). In addition, G-banding patterns in G_2 prematurely condensed chromosomes and in the corresponding metaphase chromosomes reveal overall similarity, except that there are at least three times the number of bands in the former (Fig. 12B). It appears that many single C bands in metaphase chromosomes actually consist of a number of bands, which are visually separated in the much elongated G_2 prematurely condensed chromosomes (Unakul *et al.*, 1973). Q-banding of HeLa G_2 prematurely condensed chromosomes has also been detected, although the patterns are not yet known (Patil *et al.*, 1972). G_1 banding can be seen by either Giemsa or quinacrine staining (Unakul *et al.*, 1973), but the detailed correlation in banding between prematurely condensed chromosomes of either G_1 or G_2 and metaphase chromosomes awaits further investigation. Nevertheless, present data strongly suggest that the induction of PCC in both G_1 and G_2 nuclei corresponds to a normal condensation of interphase chromatin into the individual chromosomal elements which that cell would have formed at metaphase.

Under the electron microscope after sectioning, prematurely condensed chromatin appears as "less condensed chromatin with a variable morphology ranging from hardly detectable thin threads of chromatin granules to more compact masses of chromatin closely resembling the chromatin seen in a normal prophase nucleus" (Aula, 1970). "The pulverized material consists of fuzzy masses of various sizes—less dense than metaphase chromosomes and is devoid of any details and definite fine structure" (Sanbe *et al.*, 1970). These studies by Aula (1970) and Sanbe *et al.* (1970) deal with random cell populations; particularly in the former case, it is possible that G_1 and G_2 types of PCC were being observed. These probably constituted the "more prophaselike" material. Both articles reported that, in rare instances, microtubules were seen in close

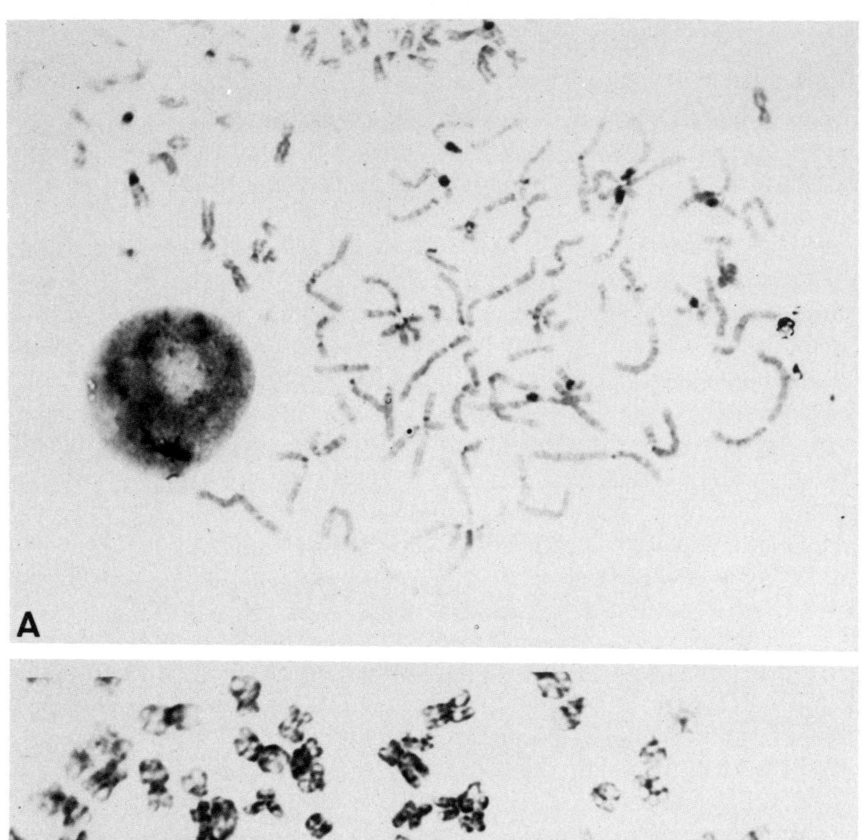

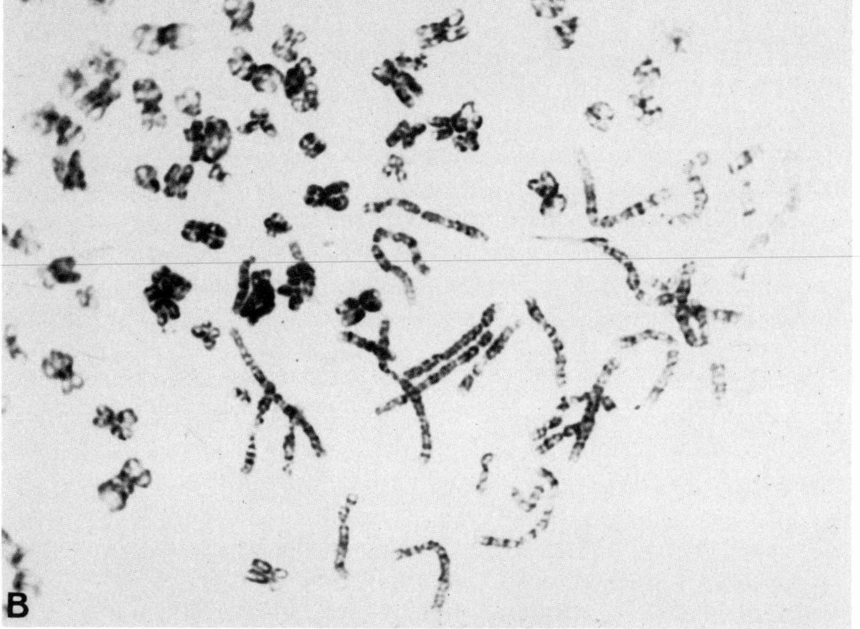

FIG. 12. Chromosomal banding patterns in G_2 prematurely condensed (PC) chromosomes. (A) C-bands are seen as black dots on G_2 PCC. (B) G-banding pattern in G_2 PCC. Usually the number of bands seen in G_2 PCC are more numerous than those in metaphase chromosomes. [From Unakul *et al.* (1973). *Nature* (*London*) *New Biol.* **242**, 106.

contact with pulverized chromatin; however, Aula (1970) stated that kinetochores were not seen in the loose type of pulverization (i.e., probably S-PCC). By contrast, Matsui *et al.* (1972) observed a kinetechore region in one area of prematurely condensed material.

C. Causes of Premature Chromosome Condensation

Some confusion has arisen in the literature over the past few years concerning the terms "chromosome pulverization" and "premature chromosome condensation." Most of this confusion centers around the findings that gross aberrations in chromosome coiling, or massive fragmentation of chromosomal material, frequently are associated with multinucleate cells formed in the presence of myxoviruses. The question arises: Is chromosome fragmentation in virus-induced multinucleate cells the result of virus damage to the genetic material, or is it due to inherent asynchrony within the multinucleate cell involving the various nuclei and cytoplasms in PCC?

It may be helpful to document some of the findings that relate to these points: (1) Virus-induced chromosome damage is allegedly of two kinds: (a) single chromatid breaks or delayed isolocus breaks (Östergren and Wakonig, 1954); and (b) massive fragmentation or pulverization (Nichols *et al.*, 1965). (2) Single chromatid breaks in virus-infected material are commonly observed in mononucleate cells (Fig. 13A), while pulverization (Fig. 13B) is most frequently associated with multinucleate cells (Nichols *et al.*, 1964, 1965; O'Neill and Miles, 1969). (3) Most viruses belonging to the NDV group of myxoviruses (Waterson, 1962) are associated with chromosome pulverization. These viruses, including measles, Sendai, and mumps, also promote widespread cell fusion (Roizmann, 1962). Another myxovirus, influenza, does not cause fusion and is not associated with pulverization (Cantell *et al.*, 1966). (4) Other groups of viruses which also cause cell fusion accompanied by chromosome pulverization include herpes (Benyesh-Melnick *et al.*, 1964) and adenoviruses (Roizmann, 1962). (5) Complete chromosome pulverization is found only rarely in mononucleate cells (Harnden, 1964), but where it does occur, only a few chromosomes are usually involved (Zur Hausen, 1967; Miles and O'Neill, 1969; Nichols *et al.*, 1965; Stenman and Saksela, 1969). Generally these chromosomes are the last to complete replication (Stenman and Saksela, 1969). (6) It is also clear that the incidence of chromosome pulverization increases in mononucleate cells as the ploidy of the cell line increases (Miles and O'Neill, 1969). (7) When multinucleate cells are produced by means

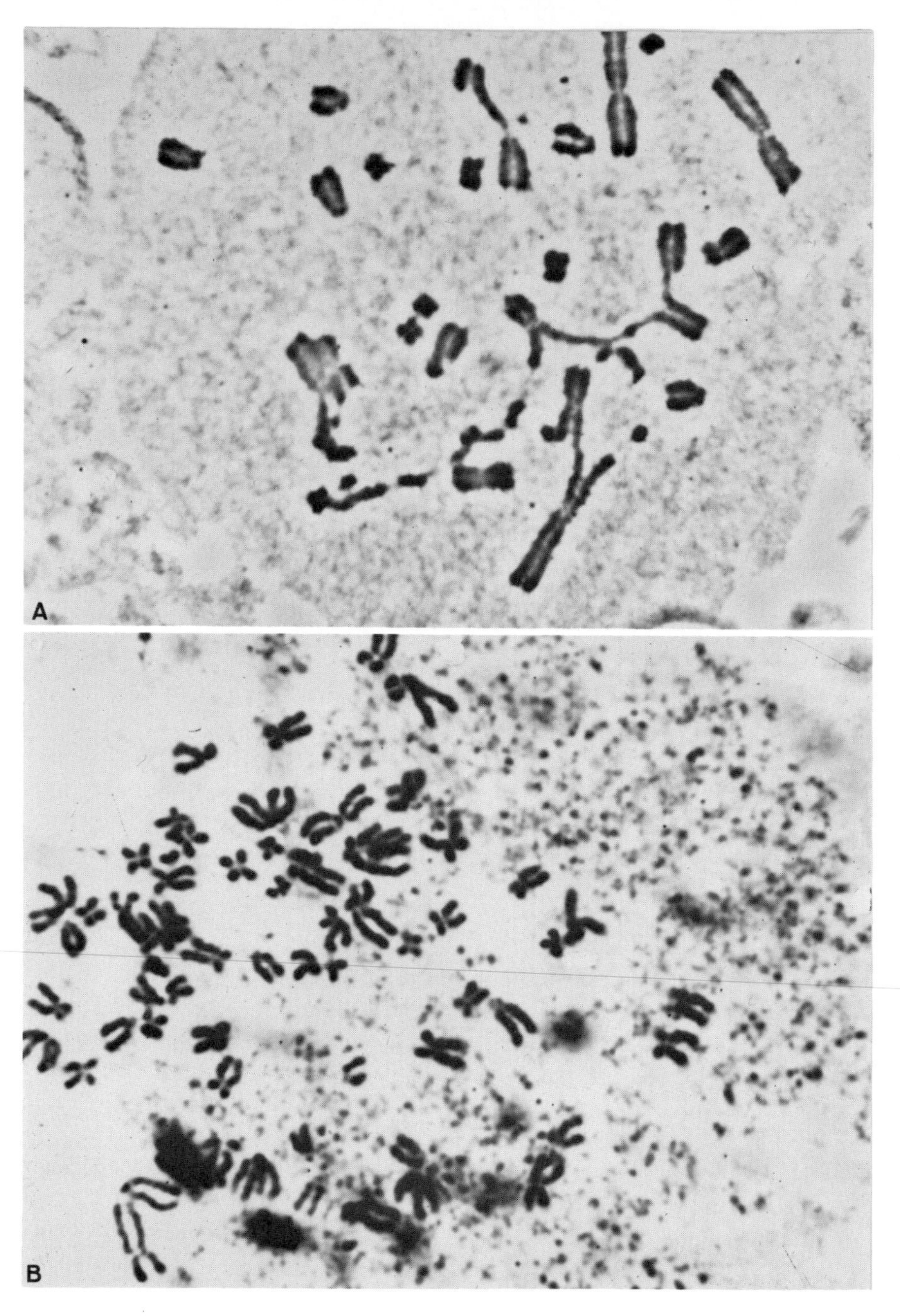

FIG. 13. Viral damage to cells. (A) Chromatid breaks in Chinese hamster (DON) cells exposed to Adeno-12 virus for 24 hours. [Figure by courtesy of Drs. Hans F. Stich and T. C. Hsu.) (B) A syncytium with intact and pulverized chromosomes after treatment with hemolytic fraction of measles virus. [From Nichols *et al.*, (1967). *J. Cell Biol.* **35**, 257.]

of UV-inactivated Sendai virus from parental cells occupying G_2 and S stages in the cell cycle, a small proportion of these cells as they pass into mitosis show aberrations of chromosome condensation in the S nuclei. These S nuclei were still involved in DNA replication at the time of aberrant condensation (Johnson and Rao, 1970). Various other investigators have associated aberrant condensation (or pulverization) in multinucleate cells with a late-replicating nucleus (Nichols *et al.*, 1967; Kato and Sandberg, 1967; Stenman and Saksela, 1969). (8) Aberrations in chromosome condensation occur in multinucleate cells formed as a result of exposure either to active (Miles and O'Neill, 1969) or inactive myxovirus (Johnson and Rao, 1970; Johnson *et al.*, 1970; Takagi *et al.*, 1969). (9) The type of aberration in chromosome condensation is directly related to the phase of the cell cycle which the interphase nucleus occupies at the time of cell fusion with a mitotic cell (Johnson and Rao, 1970; Stenman and Saksela, 1971).

These data allow us to differentiate between direct and indirect effects of a virus with respect to chromatid breaks and PCC. The direct effects include single chromatid breaks in mononucleate (and presumably multinucleate) cells following infection with *live* virus. Furthermore, this damage is not seen for many hours after infection, i.e., until the chromosomes become visible during mitosis. The indirect effects are a consequence of fusion between adjacent cells which are asynchronous with respect to one another in the cell cycle. This artificially generated internal asynchrony can, in some circumstances, lead to premature chromosome condensation of one or more, but not all, of the component nuclei (Johnson and Rao, 1972). In addition, this type of chromosomal aberration is found *immediately after* the fusion event, and occurs with equal frequency whether the virus is active or inactive.

Other considerations lead us to conclude that it is the multinucleate condition rather than the viral infection which leads to PCC. For example, Sendai virus which has been inactivated either by UV (Johnson and Rao, 1970) or by β-propiolactone (Stenman and Saksela, 1971) has no infectious properties, but retains the ability to fuse cells; this leads to a multinucleate condition, in which the three types of prematurely condensed chromosomes observed are identical to those found when live virus is used. Moreover, the kinetics of appearance of PCC are identical whether live or inactive virus is the agent of fusion. Premature chromosome condensation is almost always associated with multinucleate cells derived from fusion between G_2 and S phase cells, or between mitotic cells and cells in any other part of the cell cycle. The fact that PCC is observed only in G_2/S or mitotic interphase fusions, but not in G_1/G_2 or G_1/S fusions, implies that PCC is the result of

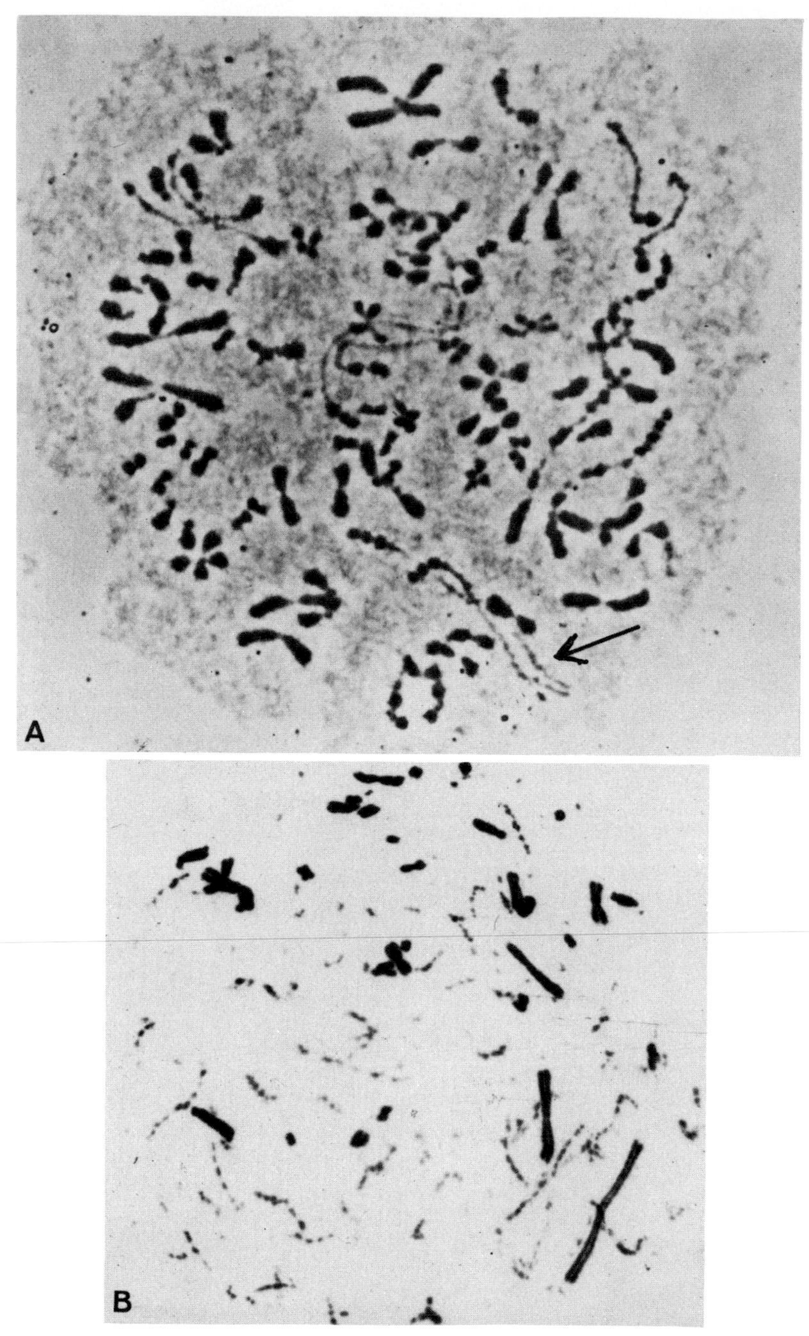

FIG. 14. Aberrant chromosome condensation in mononucleate cells. (A) Octaploid DON-C cell containing normal and aberrantly condensed chromosomes. (Figure by courtesy of Dr. Elton Stubblefield.) (B) Chromosomal pattern observed in a random culture of human–Chinese hamster cell hybrids sampled after 2 days of

an interaction between cells in specific phases of the cell cycle (Johnson and Rao, 1970).

The occurrence of PCC in several permanent cell lines which are predominantly mononucleate and have not been exposed to myxovirus is well established (Zur Hausen, 1967; Miles and O'Neill, 1969). In these lines the incidence of PCC is always low and is restricted to a few chromosomes (Fig. 14A), which are usually late replicating with regard to the rest of the genome (Kato and Sandberg, 1968a; Stenman and Saksela, 1969). In polyploid strains of these cell types, the incidence of PCC is much higher (Miles and O'Neill, 1969). We believe that premature condensation of a complete S phase nucleus in a multinucleate cell, and the aberrant coiling or pulverization of a few chromosomes in diploid and polyploid mononucleate cells, have their origins in a similar mechanism, i.e., the asynchrony among different components in their entry into mitosis. Studies with multinucleate cells support this hypothesis (Stubblefield, 1964; Kato and Sandberg, 1968b,c; Miles and O'Neill, 1969; Johnson and Rao, 1970; Stenman and Saksela, 1969), and there is also some evidence that in mononucleate cells, those chromosomes which tend to become either wholly or partially pulverized replicate later than the others (Stenman and Saksela, 1969; Miles and O'Neill, 1969). The increased incidence of PCC in polyploid cell lines is probably due to a greater degree of asynchrony in the completion of DNA synthesis by the various chromosomes. In a few cells, the majority of the chromosomes enter mitosis while one or two chromosomes are still in S phase, and this results in the PCC of these late-replicating elements. A similar situation was reported by Kao and Puck (1970) in human–Chinese hamster cell hybrids, where some of the human chromosomes were prematurely condensed (Fig. 14B).

V. The Fate and Consequences of Premature Chromosome Condensation

A. Template Activity of Prematurely Condensed Chromosomes

It is well established that both metaphase chromosomes and heterochromatic regions of the genome have much reduced template activity (Brown, 1966). Evidence that prematurely condensed chromosomes con-

---

growth following fusion. All or most of the Chinese hamster and approximately ten human chromosomes are mitotic. The remaining chromosomes display a "pulverized" appearance reminiscent of that obtained when a mitotic cell is fused with a cell in S. [From Kao and Puck (1970). *Nature* **228**, 329.]

tinue to synthesize RNA but at reduced levels when compared with interphase chromatin, has been provided by the work of Aya and Sandberg (1971) and of Stenman (1971). The amount of RNA synthesized by PCC decreases with time after fusion (Aya and Sandberg, 1971); this suggests that the templates become increasingly inaccessible to the enzymes and cofactors concerned with transcription as induction progresses. Stenman (1971) also reported that the synthesis of RNA by G_1 (intermediate) PCC was inhibited earlier than either S (pulverized) or G_2 (prometaphase-like) forms of PCC.

Johnson and Rao (1970) found that S phase prematurely condensed chromosomes continue to synthesize DNA for up to 75 minutes after the completion of fusion (Table II). As compared with corresponding nonfused or fused S phase cells in the same population, the amount of DNA synthesized by S phase prematurely condensed chromosomes was much reduced (based on grain counting in autoradiographs). Two facts emerged clearly: first, the extent of incorporation of thymidine-^{3}H into the DNA of S phase prematurely condensed chromosomes is related to the time of the induction, i.e., less DNA was synthesized at later periods of incubation. Second, the incorporation of thymidine-^{3}H into DNA is related to the ratio of mitotic versus S phase nuclei in the cell at the time of fusion; the greater the proportion of mitotic elements,

TABLE II
DNA SYNTHESIS IN PREMATURELY CONDENSED (PC) S NUCLEI[a]

	Time (min after addition of virus)	PCC scored				
		3M/1S	2M/1S	1M/1S	1M/2S	Total
Percent incidence of DNA synthesis in PC nuclei[b]	75	20.0(5)	45.0(20)	70.6(48)	100.0(21)	69.0(94)
	195	0 (7)	11.0(18)	44.4(54)	77.8 (9)	37.0(88)
Average number of grains per PC nucleus[c]	75	2.0(1)	3.1 (9)	4.8(34)	4.5(42)	4.3(88)
	195	0(0)	1.0 (2)	3.2(24)	3.5(14)	3.2(40)

[a] From Johnson and Rao (1970). *Nature* **226**:717, with permission.
[b] Number of cells scored for each class is given in parentheses.
[c] Number of PC nuclei on which grains were counted is given in parentheses.

the less was the incorporation of thymidine-^{3}H into S phase prematurely condensed chromosomes, and the sooner it was extinguished. These data again point to an increasing inaccessibility of template DNA when the PCC induction process occurs. In this example, it is DNA synthesis which is turned off.

B. Integration or Exclusion of Prematurely Condensed Chromosomes in Hybrid Cells

Two of the leading questions about prematurely condensed chromosomes concern (1) whether they are retained in the progeny of the M/I fused cells; and (2) if so, are they genetically active?

At present there is little information about the coupled induction of chromosome condensation and spindle formation. Studies at both the light microscope level (Takagi *et al.*, 1969; Rao and Johnson, 1972a; Matsui *et al.*, 1972) and the electron microscope level (Aula, 1970; Sanbe *et al.*, 1970; Matsui *et al.*, 1972) show that the induced chromosomes usually remain grouped together in one or more coherent masses. Rao and Johnson (1972a) reported that prematurely condensed chromosomes of G_1, S, and G_2 types were in some way spatially associated with the metaphase chromosomes of the inducer cell, even though it was impossible to define any precise arrangement on the spindle (Fig. 15). In a few instances, microtubules have been found associated with prematurely condensed chromosomes (Aula, 1970; Sanbe *et al.*, 1970), but in only one example has a kinetochore been seen in prematurely condensed chromosomes (Matsui *et al.*, 1972). Nevertheless, it is possible that there is some centromeric differentiation in both G_2 and G_1 chromosomes, since C-banding has been demonstrated in these interphase chromosomes (Unakul *et al.*, 1973).

An association between prematurely condensed chromosomes and the spindle has been described both in the endosperm of *Haemanthus* (Östergren and Bajer, 1961) and in nuclear transplantation experiments which involved the introduction of nuclei into maturing oocytes of *Xenopus* (Gurdon, 1968). In the first example, a few nuclei of the syncytial tissue were observed to enter mitosis simultaneously with other nuclei, but these contained elongated single chromatids rather than double metaphase chromosomes. Such single chromatids appeared to have great difficulty in establishing clear metaphase plates, and in many instances they remained outside the equatorial plate of the spindle; however, a number of these chromatids did arrange themselves on the metaphase plate. Segregation of the single chromatids at anaphase was com-

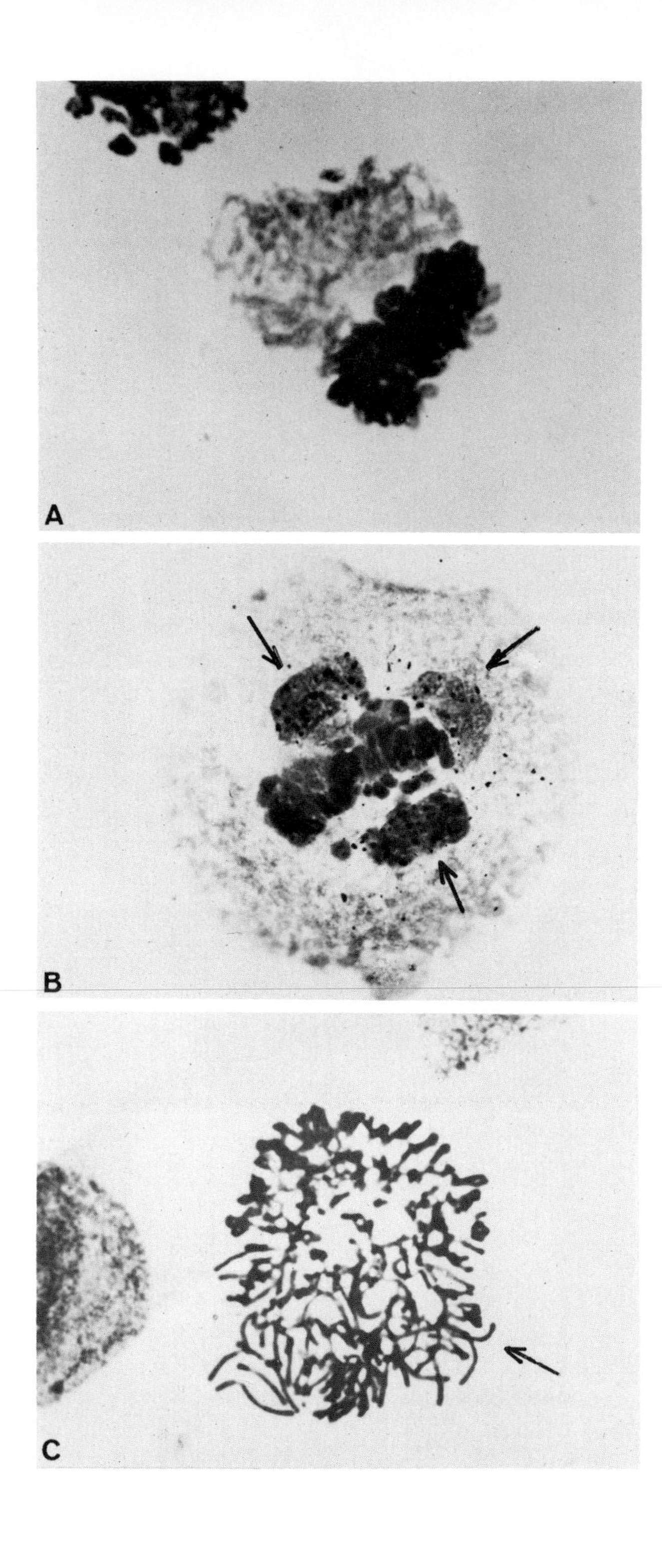
A
B
C

plex: some elements moved to one or the other pole, but many lagged behind. Nevertheless, there were indications in this living tissue that the single chromatids which did move to the poles may have done so due to an active association between the centromeric region and the spindle. The lagging chromosomes formed telophase nuclei which were often deformed and torn apart. Östergren (1961) has discussed the theoretical aspects of these observations made on *Haemanthus* endosperm, and has also reviewed the earlier literature. In the nuclear transplant studies of Gurdon (1968), induction of chromosome condensation in nuclei transplanted into maturing (i.e., meiotic) oocytes was associated with the development of multipolar spindles and numerous asters. Although the induced chromosomes were clearly arranged on the spindles and in some cases on the equator (see Gurdon, 1968, Figs. G–J), in general the arrangement was irregular and the chromosomes did not segregate evenly.

These various observations allow us to draw two important conclusions. The first is that, when PCC is induced in *Haemanthus* endosperm, it is accompanied by the induction of a spindle. The second conclusion is that the induced chromosomes are clearly associated with the spindle, although the aberrant segregation patterns imply that association is not as well organized as in the normal mitotic or meiotic processes.

In order to assess the segregation of induced chromosomes, Rao and Johnson (1972a) brought about fusion between reversibly blocked mitotic HeLa cells and synchronized populations of G_1, S, and G_2 HeLa cells. Completion of mitosis in these cells did not occur for up to 7 hours; therefore it was possible to examine the patterns of segregation very closely. It was observed that the prematurely condensed chromosomes of each phase very quickly adopt a position adjacent to the metaphase chromosomes (Fig. 15); as anaphase begins, prematurely condensed chromosomes of the S type tend to become more scattered throughout the cell than those of either G_1 or G_2 type (Fig. 16). Final segregation patterns are very variable. For example, the prematurely condensed chromosomes could pass entirely into one of the two daughter

---

FIG. 15. Premature chromosome condensation (PCC) in M/G_1, M/S, and M/G_2 fusions of HeLa cells at 30 minutes after fusion. Mitotic cells were obtained by application of N_2O block. Interphase cells were prelabeled with thymidine-^3H prior to cell fusion. (A) PCC of the G_1 type which is lightly stained and lies alongside the metaphase plate (darkly stained) of the mitotic cell. (B) Autoradiograph showing labeled S type PCC (indicated by arrows) lying around the metaphase plate. (C) Autoradiograph showing labeled PCC of the G_2 type (indicated by the arrow) lying adjacent to the metaphase chromosomes of a mitotic cell. [From Rao and Johnson (1972a). *J. Cell Sci.* **10**, 495.]

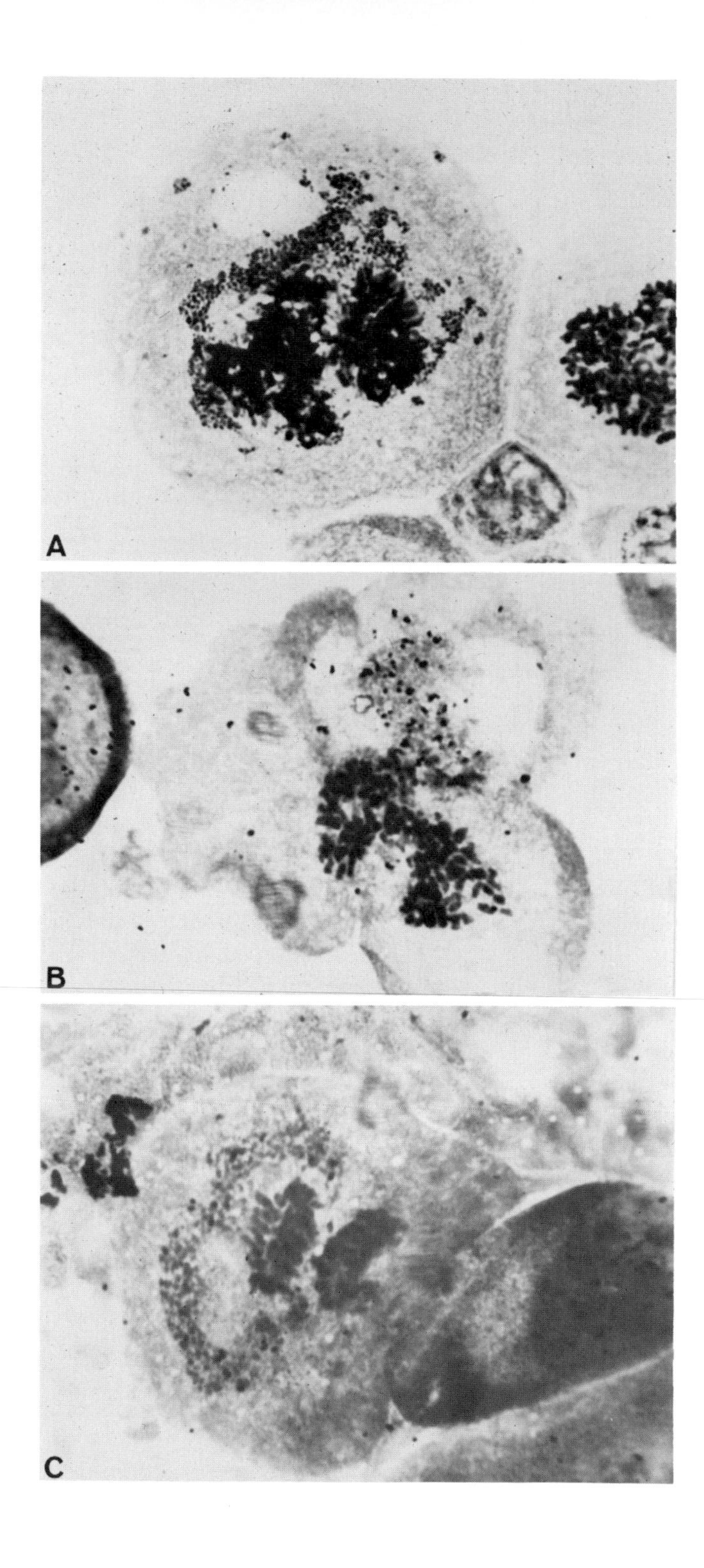
A
B
C

cells (Fig. 16C), or might pass unequally into both daughters (Fig. 16A). Alternatively, they could segregate into a daughter cell which did not receive any of the mitotic component (Fig. 16B). Completion of mitosis at telophase could result in the incorporation of all the prematurely condensed chromosomes into one or both daughter nuclei (Fig. 17A). Those that are incorporated into the daughter nuclei of a hybrid cell go through the mitotic cycle and reappear as normal metaphase chromosomes during the second mitosis after fusion (Fig. 17B). These findings, and those of Takagi *et al.* (1969), confirm that prematurely condensed chromosomes do indeed pass into the progeny of M/I fused cells, and that the pattern of segregation is exceedingly complex. Such complexity in segregation strongly suggests that the process of induced condensation is not accompanied by the formation of a functional spindle in the interphase cell; however, neither does it rule out the possibility of incipient spindle development.

The genetic implications of carrying over prematurely condensed chromosomes of G_1 and S types into daughter cells have been alluded to by Johnson and Rao (1972); this may constitute one way of introducing genetic variability into a population of mononucleate cells. In a series of fusion experiments, Johnson and Rao (unpublished) observed that prematurely condensed chromosomes were carried over into the next cycle following completion of mitosis, and that they still appeared at the next mitosis (Fig. 17C). In the first mitosis following fusion between reversibly blocked mitotic HeLa cells and interphase HeLa cells, some of the chromosome spreads contained either G_1 or S prematurely condensed chromosomes. By the second mitosis after fusion, the occurrence of these few prematurely condensed chromosomes was very low, and by the third mitosis they had disappeared. Is it possible for the few G_1 prematurely condensed chromosomes to be carried over intact within the nucleus of a daughter cell? This seems unlikely, since they would almost certainly replicate and appear as G_2 or normal chromosomes. More likely is the possibility that they are carried over in a micronucleus,

FIG. 16. Scattering of prematurely condensed (PC) chromosomes in M/S fused HeLa cells at 4 hours after fusion. S-phase cells prelabeled with thymidine-^{3}H were fused with mitotic cells reversibly blocked with N_2O. (A) The mitotic chromosomes are completing anaphase while the PC chromosomes are scattered around them. PC chromosomes will probably be distributed randomly among the daughter nuclei. (B) Autoradiograph of an M/S fused cell during anaphase, with the labeled PC chromosomes localized to one side of the dividing cell. (C) In this M/S fused cell, the PC chromosomes form a ring around one of the anaphase chromosome groups, and in all probability will become part of that daughter cell. [From Rao and Johnson (1972a). *J. Cell Sci.* **10**, 495.]

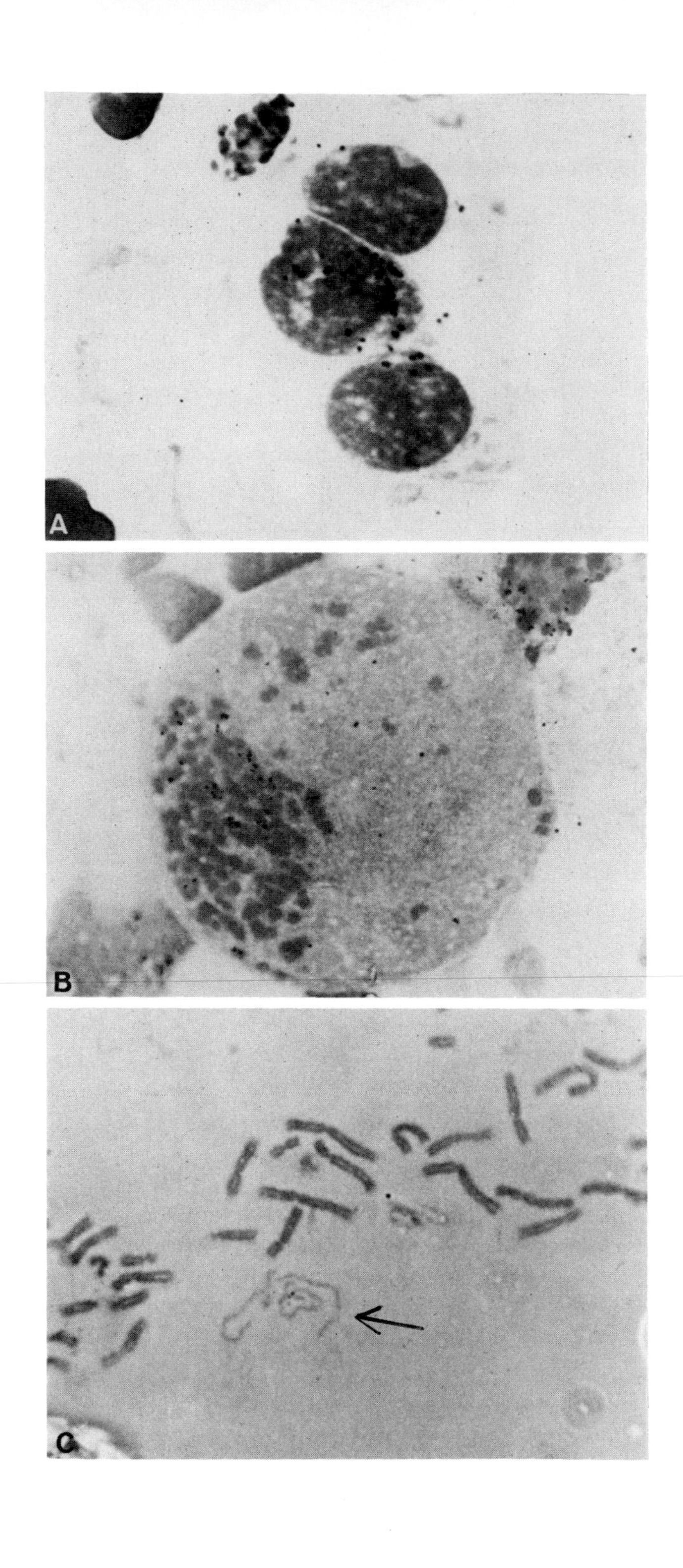
A
B
C

which for some reason does not synthesize DNA (possibly because of the absence of a nucleolar region; see Das, 1962).

The question of whether prematurely condensed chromosomes can be retained indefinitely in the genome, and more particularly whether chromosome fragments can become stably integrated by mechanisms of recombination or translocation into another intact chromosome is a critical issue. Rao and Johnson (1972a) attempted to answer the first point by fusing synchronized populations of two complementing auxotrophic mutant lines of Chinese hamster ovary cells. By examining the plating efficiencies of the various mitotic-interphase and interphase-interphase fusions in selective medium, it was possible to assess whether prematurely condensed chromosomes were carried over into the progeny (Table III). The fact that the plating efficiency in the mitotic interphase fusions is only about 50% of the value for interphase/interphase fusions indicates that the induction of PCC reduces plating efficiency; this may be due to the random distribution of prematurely condensed chromosomes during the cell division immediately following fusion. Relatively higher plating efficiencies for M/G_1 and M/G_2 fusions, in comparison to M/S fusion, indicate that the G_1 or G_2 prematurely condensed chromosomes have a better chance of being retained by the hybrid cell than S-phase ones. This is to be expected in view of the uneven condensation of S-phase prematurely condensed chromosomes and the problems involved in segregation of such material. Our conclusion is that, in mitotic-interphase fused cells, prematurely condensed chromosomes are often incorporated into the daughter nuclei of the hybrids and are functionally retained by the progeny.

Stable integration of chromatin fragments produced by PCC into the genome of another cell has also been postulated to account for the "genetic repair" found after A-9 mouse fibroblasts deficient in the enzyme *hypoxanthine guanosine phosphoribosyl transferase* (HGPRT) were fused with embryonic chick erythrocytes (Schwartz *et al.*, 1971). In this case it was impossible to detect the chick genetic material.

FIG. 17. Incorporation of prematurely condensed chromosomes (PCC) into the daughter nuclei of hybrid cells. (A) M/G_1 fused cell at the end of the first mitosis after fusion. The G_1 PCC are incorporated into 2 of the 3 daughter nuclei, as indicated by the presence of label. (B) M/G_2 hybrid cell during the second mitosis after fusion, showing labeled and unlabeled chromosomes on the same metaphase plate. The labeled chromosomes were derived from the incorporation of labeled PCC of the G_2 type. (C) Appearance of PCC during second mitosis following cell fusion in Chinese hamster ovary cells. Only a fraction of the genome (probably three chromosomes indicated by the arrow) is prematurely condensed. [From Rao and Johnson (1972a). *J. Cell Sci.* **10**, 495.]

TABLE III
RATE OF SURVIVAL OF HYBRIDS IN HOMO- AND HETEROPHASIC FUSIONS BETWEEN SYNCHRONIZED POPULATIONS OF GLY-A AND GLY-B MUTANTS[a]

		Type of fusion[b]	No. of colonies per 1000 cells of each parental type	Relative plating efficiency (%)
Random				
	1	R^a/R^b	15.6	—
	2	R_c^a/R_c^b	5.0	109.0
	3	$R^a + R^b$ (w/o virus)	0.0	
Homophasic				
	4	M^a/M^b	4.0	—
	5	G_1^a/G_1^b	5.5	—
	6	S^a/S^b	4.6	—
	7	G_2^a/G_2^b	4.4	—
			Average 18.5/4 = 4.6	100.0
Heterophasic				
(i) Mitosis/interphase				
	8	M^a/G_1^b	2.0	43.5
	9	M^a/S^b	0.62	13.5
	10	M^a/G_2^b	1.56	34.0
(ii) Interphase/interphase				
	11	S^a/G_1^b	3.2	—
	12	S^a/G_2^b	2.7	—
	13	G_1^a/G_2^b	2.5	—
			Average 8.4/3 = 2.8	61.0

[a] From Rao and Johnson (1972a). *J. Cell Sci.* **10**:495, with permission.
[b] R, Random population; [a], parent Gly-A; [b], parent Gly-B; [c], exposed to Colcemid (0.05 μg/ml) for 2 hours before fusion.

C. EFFECT OF AN INTERPHASE COMPONENT ON METAPHASE CHROMOSOMES

In multinucleate cells containing a high ratio of interphase to mitotic nuclei, it is improbable that any of the interphase nuclei will be induced into PCC, even though some slight condensation frequently can be seen around the nuclear envelope. Ikeuchi *et al.* (1971) and Rao and Johnson (1972a) reported that one of the alternatives to induction of PCC in this system is a pycnosis of the mitotic chromosomes, which come together as a tightly packed ball (Fig. 18). The greater the I:M ratio, the greater is the chance that this will occur (Ikeuchi *et al.*, 1971); in addition, pycnosis of the mitotic chromosomes is related to the length of time which the cells spend in the presence of Colcemid. There appears

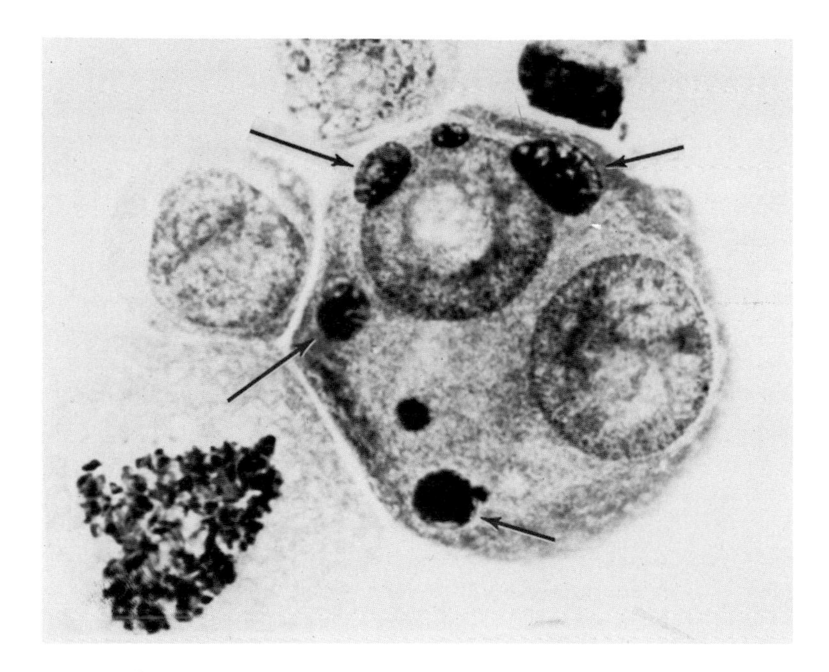

FIG. 18. Effect of an interphase component on metaphase chromosomes. M/S fused cell in which there was no induction of premature chromosome condensation in the S nuclei. The mitotic chromosomes were transformed into micronuclei under the influence of the interphase nuclei. Micronuclei are darkly stained.

to be a substantial decay or breakdown of the cellular mechanisms responsible for maintenance of the mitotic state (Ikeuchi *et al.*, 1971; Rao and Johnson, 1972a). In DON Chinese hamster cells, the pycnotic masses of chromatin (which have been designated "telophase-like-nuclei" by Ikeuchi *et al.*, 1971), are very rapidly enclosed in a nuclear envelope. These micronuclear masses are sometimes passed on to the daughter cells, but they remain inactive in template function, synthesizing neither RNA nor DNA (Rao and Johnson, 1972a).

These observations suggest that maintenance of the mitotic state depends upon a subtle balance of materials which ensure that the chromosomes will remain discrete until they segregate. Possibly these materials are diluted out in multinucleate cells, or else depleted by aging the mitotic population in Colcemid for long periods before fusion takes place. However, it should be noted that interphase nuclei on rare occasions may remain unaffected in the presence of a mitotic component, and in this state may pass into one or none of the daughter cells, along with the reconstituted telophase nuclei.

VI. Factors Involved in the Induction of Premature Chromosome Condensation

In Section III, we discussed attempts to induce chromosome formation either in intact cells or in isolated nuclei by a variety of treatments involving hypertonicity, lowering of pH, or adding basic molecules. The conclusions reached were: (1) prophaselike condensation patterns can be achieved by some treatments, and these are reversible in some cases; but (2) no true chromosome condensation has yet been achieved, probably because some other factors are missing. Prophaselike inductions by chemical agents mimic but do not absolutely correspond to the pattern of true prophase organization of chromatin (Philpot and Stanier, 1957). By contrast, it seems clear that when mitotic and interphase cells are fused, a true chromosome condensation occurs in the interphase nucleus, and that this is directly related to the ratio of mitotic to interphase elements in the cell. An important question can be posed, therefore: What factors exist in the mitotic cell which produce this ordered condensation?

A. Effect of Anions and Cations in the Induction of Premature Chromosome Condensation

Since a variety of cations promote some degree of condensation in intact cells and isolated nuclei, Rao and Johnson (1971) carried out a series of studies with the "mitotic-interphase fused cell system" (MIFCS) to examine the promotion or suppression of chromosome condensation when cations or anions were present at elevated concentrations in the medium. Table IV summarizes the effects of various positively charged compounds on the frequency of PCC induction. Among the polyamines, spermine and putrescine were found to promote PCC induction, while surprisingly, spermidine inhibited this process. Addition of cadaverine had no effect. Of the two divalent cations tested, Mg^{2+} promoted PCC induction whereas Ca^{2+} did not. These data suggest that there is some degree of specificity in the effect of cations on PCC induction, and this may be related to the ability of these molecules to form stereospecific complexes with chromatin.

The addition of anions, or negatively charged compounds, to a fusion mixture definitely inhibits the induction of PCC in this system (Table V). Unlike positively charged compounds, the negatively charged ones

TABLE IV
EFFECT OF POLYAMINES AND DIVALENT CATIONS ON THE INDUCTION OF PREMATURE CHROMOSOME CONDENSATION (PCC) IN THE MITOTIC-INTERPHASE FUSED CELL SYSTEM (MIFCS)[a]

		Incidence of PCC(%) among hybrid cells[b]			
		Binucleate		Trinucleate	
Treatment	Dose	1M:1I	% PCC relative to control	2M:1I	1M:2I
Control		50.0	100.0	79.4	12.2
Putrescine	2×10^{-3} *M*	67.0	134.0	80.0	33.4
Spermine	2×10^{-3} *M*	81.0	162.0	100.0	76.0
Cadaverine	2×10^{-3} *M*	47.5	95.0	76.2	0
Spermidine	2×10^{-3} *M*	14.0	28.0	14.4	0
$MgCl_2$	2×10^{-3} *M*	65.0	130.0	95.0	40.0
$CaCl_2$	2×10^{-3} *M*	42.5	85.0	70.0	45.0

[a] From Rao and Johnson (1971). *J. Cell Physiol.* **78**:217 with permission.
[b] M, mitosis; I, interphase.

TABLE V
INHIBITORS OF PREMATURE CHROMOSOME CONDENSATION (PCC) INDUCTION IN MITOTIC-INTERPHASE FUSED CELL SYSTEM (MIFCS)[a]

		Incidence of PCC (%) among hybrid cells[b]			
		Binucleate		Trinucleate	
Treatment	Dose	1M:1I	% PCC relative control	2M:1I	1M:2I
Control		50.0	100.0	79.4	12.2
Estradiol-17β	8×10^{-5} *M*	14.0	28.0	14.4	0
$Na_2H\ PO_4$	2×10^{-2} *M*	15.0	30.0	31.0	0
EDTA	0.1%	19.7	39.4	22.6	0
Cyclic AMP	2×10^{-3} *M*	22.4	44.8	29.0	0
Heparin	100 units/ml	26.4	52.8	45.5	0

[a] From Rao and Johnson (1971). *J. Cell Physiol.* **78**:217, with permission.
[b] M, mitosis; I, interphase.

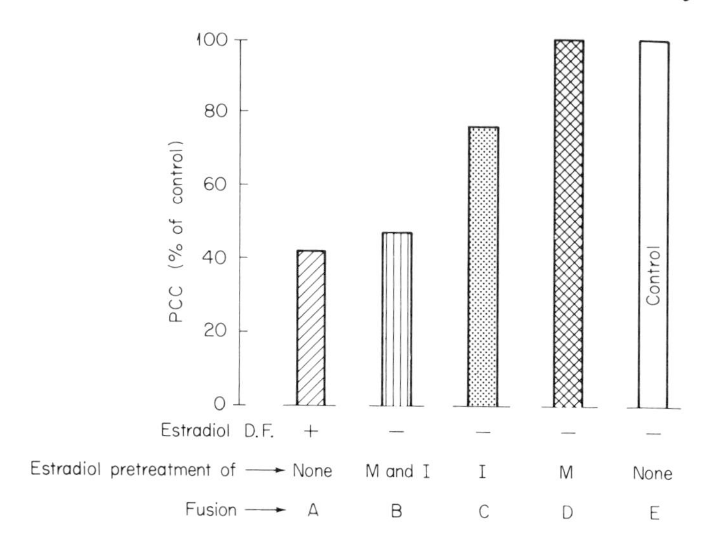

FIG. 19. The site of action of estradiol-17β in blocking induction of premature chromosome condensation (PCC). The presence of estradiol (80 m*M*) during fusion (D.F.) is indicated by +, and its absence by —. The type of cell population that received estradiol pretreatment is also listed under each column. M, mitotic cells; I, interphase cells. [From Rao and Johnson (1971). *J. Cell Physiol.* 78, 217.]

are uniform in their effects, i.e., they always inhibit PCC induction, even in cells containing two mitotic to one interphase component. However, the inhibitory effect of 17β-estradiol cannot be explained by charge effects, since at the prevailing pH (7.0) this molecule is not negatively charged; nevertheless, it produced a greater reduction in the frequency of PCC in the fused cells than any of the negatively charged compounds used in the study.

In order to investigate how 17β-estradiol blocks PCC, and more particularly whether it exerts its effect on the interphase or mitotic component of the system, five different fusions were carried out between mitotic and interphase HeLa cells. To fusion A, 17β-estradiol was added along with virus. To fusion E (control), no estradiol was added. In fusion B, both mitotic and interphase cells were treated with estradiol for 30 minutes before fusion, then washed before the addition of virus. In C only the interphase cells, and in D only the mitotic cells, were treated with estradiol before fusion. Figure 19 shows that there was little difference in the incidence of PCC when both mitotic and interphase cells were exposed to estradiol, either during or before fusion. However, inhibition of PCC occurred when the interphase population alone was pretreated; pretreatment of the mitotic cells had no effect on the frequency of PCC. The absence of any change in the ability of estradiol-pretreated mitotic cells to induce PCC is somewhat surprising, but in view of the fact that estradiol induces a G_2 delay in HeLa cells (Rao, 1969), it is possible that the interphase population treated with estradiol was rendered relatively insensitive to the mitotic inducer components. In this case it seems unlikely that estradiol binds to the inducer molecules, whatever they may be, but it may exert its effect on the interphase chromatin or at the nuclear envelope. This inhibitory action of 17β-estradiol can be nullified if a polycation, such as putrescine dihydrochloride, is added simultaneously to the fusion mixture.

B. Involvement of Proteins in the Induction of Premature Chromosome Condensation

The induction of complete chromosomes by mitotic-interphase fusions suggests that there may be some unique molecules in this system. There are now two pieces of evidence that implicate proteins in the formation of PCC.

1. *Synthesis of G_2 Phase Proteins Essential for Promoting PCC*

Matsui *et al.* (1971), working with the DON line of Chinese hamster cells, found that if protein synthesis is blocked by puromycin or cycloheximide during the last 65 minutes of G_2, the cells enter mitosis but there is a 50% reduction in their ability to induce PCC in interphase nuclei following fusion. However, the RNA which presumably codes for this protein is synthesized not during G_2, but considerably earlier in the cycle.

2. *Migration of Protein from Mitotic Inducer Cells to Prematurely Condensed Chromosomes*

In the following experiments of Rao and Johnson (1972b), HeLa cells were incubated in media containing a mixture of tritiated amino acids, either for one complete generation or for a 3-hour period during each phase of the cell cycle. These cells were arrested in mitosis, with the mitotic index reaching at least 97%. Following mitotic arrest, the cells were fused with an unlabeled population of interphase cells in the presence of cycloheximide (25 μg/ml) in order to block protein synthesis during fusion. The fused cells were then prepared in the standard way for examination of chromosomes. Autoradiographs of the prematurely condensed chromosomes showed that, without exception, when the mitotic component had been prelabeled before fusion there was substantial migration of labeled material to the prematurely condensed chromosomes (Fig. 20). This label was removed only by digestion with Pronase, not by DNase or RNase treatment.

Protein synthesized during the G_1 period does not migrate as extensively as either S phase or G_2 phase protein; the latter migrates to prematurely condensed chromosomes more than protein synthesized during any other phase. In addition, the extent of migration by labeled protein, as measured by the number of grains in the prematurely condensed chromosomes, is directly proportional to the amount of label present in the mitotic chromosomes residing in the same cell (Rao and Johnson, 1974). These data suggest that certain of the inducing factors in mitotic cells may be proteins, which bind to the interphase chromatin of the induced nucleus and are responsible either wholly or in part for the ordered condensation of chromosomes. Of the chromosomal proteins, most if not all of the histones are metabolically stable (Byvoet, 1966; Hancock, 1969); moreover, the migratory proteins which are synthesized during G_2, and possibly during mitosis, are not histones. They may correspond to those chromosomal proteins described by Prescott and Bender (1963), which have a rapid turnover and are synthesized at one point during each cell cycle.

In this connection, it seems significant that the ratio of protein to DNA in metaphase chromosomes is much greater than in interphase chromatin from the same species. Cantor and Hearst (1966), Salzman *et al.* (1966), and Hancock (1969) found that there is a 2-fold enrichment of nonhistone protein in metaphase chromosomes. The nuclei of highly differentiated cells, such as avian erythrocytes and spermatozoa, are also protein-deficient in comparison with normal interphase nuclei from the same species (Dingman and Sporn, 1964; Dallam and Thomas, 1953); such cells are therefore even more grossly deficient in protein compared to metaphase chromosomes. In a further series of studies, we have found that massive protein migration occurs when PCC is induced in hen erythrocyte nuclei (Johnson and Rao, unpublished results). These facts strongly suggest that, in order to achieve chromosome condensation, a number of specific proteins must become associated with the interphase chromatin. Rapid involvement of proteins in the transformation of interphase chromatin into visible chromosomes during the induction of PCC may provide one means of analyzing this problem at the

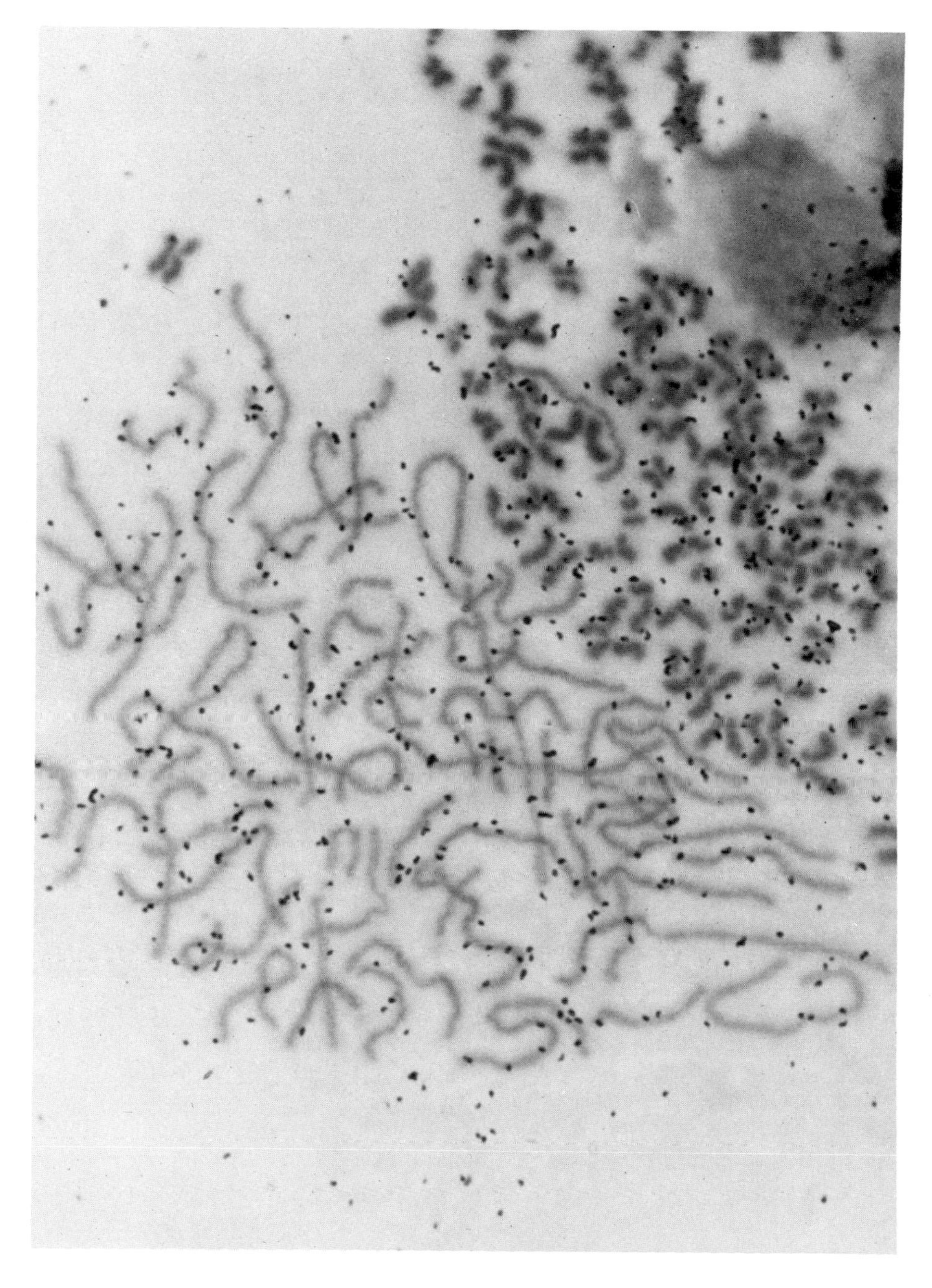

FIG. 20. Migration of labeled protein from a prelabeled mitotic cell to prematurely condensed chromosomes of an unlabeled interphase cell.

molecular level. However, further experimental evidence is necessary before we can conclude that the migration of protein to PCC is a cause of chromosome condensation.

The work of Matsui *et al.* (1971) demonstrated that protein synthesis in late G_2 appears to be involved in the capacity of a mitotic cell to induce PCC; moreover, the findings described in the preceding paragraph are compatible with the hypothesis that chromosomal nonhistone protein synthesis is essential if chromosomes are to appear. However, Matsui *et al.* (1971) proposed that three different classes of G_2 proteins are synthesized: one for the G_2-mitotic transition, another for the induction of PCC, and a third for chromosome pulverization. At present, events during the G_2 period are not well understood, except in vague terms of providing the cell with the apparatus needed for mitosis; we believe that it is not necessary in terms of present data to postulate that three different proteins are involved in chromosome condensation. Rather, we regard the pulverization and PCC-inducing factors of Matsui *et al.* (1971) as the same, and consider that these proteins are the ones which we have described in the migration studies. Other proteins synthesized during G_2 may not be connected at all with the condensation process, but with other facets of the preparation which enables an interphase cell to enter and complete mitosis.

VII. Differences between Prematurely Condensed Chromosomes and Metaphase Chromosomes

The major difference between prematurely condensed chromosomes and normal metaphase chromosomes is in the degree of condensation. Prematurely condensed chromosomes from G_1 and G_2 nuclei are not as condensed as metaphase chromosomes, and hence they do not become as intensely stained. Why is it that prematurely condensed chromosomes do not achieve the same degree of condensation as metaphase chromosomes? We may find an answer to this question by examining PCC induction in the light of a normal chromosome condensation cycle (Figs. 2 and 3). After the completion of DNA synthesis, mammalian cells usually require about 2–4 hours of preparation (G_2) before they can enter mitosis. Changes in the capacity of chromatin to bind actinomycin D (Pederson and Robbins, 1970, 1972) or safranine dye (Alvarez and Valladares, 1972), as well as changes in the sensitivity of DNA to DNase attack as a function of the cell cycle (Pederson, 1972), suggest that conformational changes or coiling take place during this time in the interphase chromatin. In this respect, G_1 chromatin is different from S or G_2 chromatin. As a matter of fact, late G_2 chromatin is probably in a different conformational state from that of early G_2 as may be seen in Fig. 3. This suggests that, after replication, the chromosomes go through some degree of maturation or coiling preparatory for mitosis. When this maturation process in the interphase nucleus is interrupted, as is the case during fusion with a mitotic cell, one consequence would be aberrant or incomplete condensation, which produces prematurely condensed chromosomes. Keeping the prematurely condensed chromosomes in proximity to metaphase chromosomes for 4 or 5 hours by Colcemid block does not cause any significant change in the length of G_1 or G_2 prematurely condensed chromosomes. Furthermore, increased mitotic:interphase ratios (e.g., 2M:1I or 3M:1I) do not seem to produce greater contraction of prematurely condensed chromosomes than that found in 1M:1I fusions. Consequently, the difference in the lengths of G_1 and G_2 prematurely condensed chromosomes probably is not due to the amount of inducer molecules available for condensation, but only reflects the degree of coiling that an interphase chromosome has gone through prior to cell fusion. The shorter the prematurely condensed chromosomes are, the greater must have been their degree of condensation during interphase.

This problem becomes even more obvious when we consider the premature condensation of one or two chromosomes, or even of a segment of a chromosome, in mononucleate cells. Usually, late-replicating chromosomes or chromosome segments become prematurely condensed when the rest of the genome enters mitosis (Stubblefield, 1966; Kato and Sandberg, 1968a). However, these prematurely condensed elements are less condensed and appear like G_2 prematurely condensed chromosomes (Fig. 13A). Possibly these segments, since they are late replicating, have not had enough time to go through the process of maturation like the rest of the genome, and hence they are poorly condensed. This phenomenon seldom occurs in mononucleate cells, but it can be induced experimentally by culturing cells in medium containing bromodeoxyuridine (Zakharov and Egolina, 1972). Chinese hamster chromosomes, and chromosomal segments which have incorporated bromodeoxyuridine into their DNA during replication, are less condensed than chromosomes that did not incorporate the analog (Fig. 21). This bromodeoxyuridine-induced "spiralization delay" (Zakharov and Egolina, 1972), which seems to be very similar to premature chromosome condensation, may be due to an altered secondary structure of the DNA molecule. Chromosomes containing bromodeoxyuridine may not have gone through the usual maturation process during G_2 because of the substitution of bromodeoxyuridine

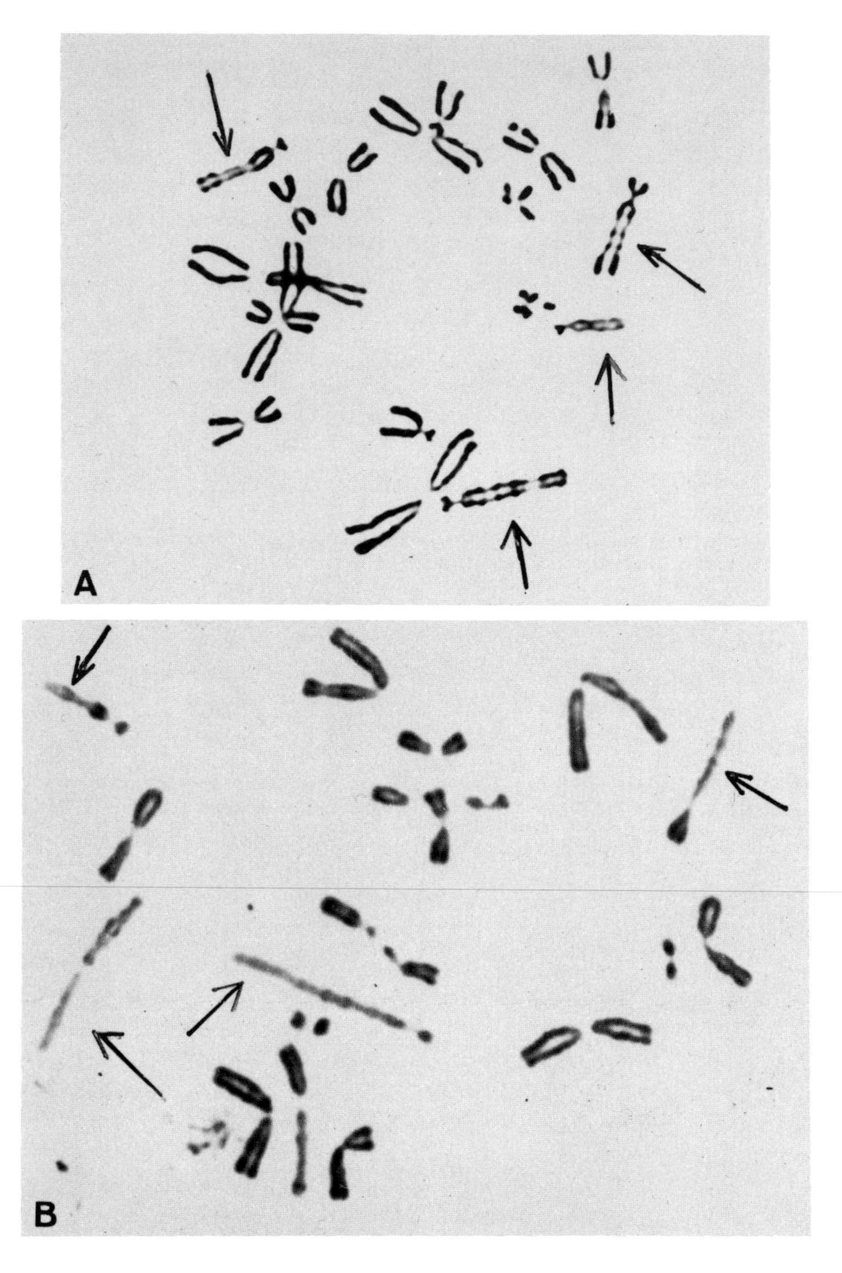

Fig. 21. Abnormal chromosome condensation (arrows) induced by bromodeoxyuridine incorporation in (A) strain 237 and (B) strain 401 of Chinese hamster cells. [From Zakharov and Egolina (1972). *Chromosoma* (Berl.) 38, 341.]

in the place of thymidine. This could then account for aberrant condensation of such chromosomes or chromosomal segments.

VIII. Conclusions

Normally the chromosomes of eukaryotic cells go through a cycle of replication, condensation, division, and decondensation. Even though chromosomes are visible only for a brief period during cell division, the diffuse chromatin in the interphase nucleus appears to undergo conformational changes which reflect the phase of the cell in the cell cycle. Attempts to induce chromosome condensation by treating interphase cells with hypertonic media or positively charged compounds, such as divalent cations and polyamines, have resulted in condensation of chromatin, but no true chromosomes have been formed. By contrast, the fusion of an interphase cell with a mitotic cell causes chromosome condensation in the former, a phenomenon which is termed premature chromosome condensation, or PCC.

Prematurely condensed chromosomes are of three different types, corresponding to the G_1, S, and G_2 stages of the interphase cell. The factors present in a mitotic cell which are responsible for chromosome condensation have yet to be defined. However, by means of the mitotic-interphase fused cell system, it has been found that Mg^{2+} among the divalent cations, and both putrescine and spermine among the polyamines, promote the induction of PCC; negatively charged compounds, on the other hand, inhibit this process. Autoradiographic evidence suggests that certain proteins are also involved in this process. Further studies are needed to isolate and characterize the chromosome condensing factors of mitotic cells. Such information is likely to be helpful in understanding the structure and organization of chromosomes.

ACKNOWLEDGMENTS

The original work presented here was supported by research Grant No. DRG-1110 from the Damon Runyon Memorial Fund for Cancer Research and the Medical Research Council of Great Britain. RTJ gratefully acknowledges the support of a Research Fellowship from Peterhouse, Cambridge.

REFERENCES

Abuelo, J. G., and Moore, D. E. (1969). *J. Cell Biol.* **41**: 73.
Adolph, E. F. (1929). *J. Exp. Zool.* **6**: 213.

Alvarez, Y., and Valladares, Y. (1972). *Nature (London), New Biol.* **238:** 279.
Anderson, N. G. (1956). *Quart. Rev. Biol.* **31:** 169.
Anderson, N. G., and Norris, C. B. (1960). *Exp. Cell Res.* **19:** 605.
Anderson, N. G., and Wilbur, K. M. (1952). *J. Gen. Physiol.* **35:** 781.
Arrighi, F. E., and Hsu, T. C. (1971). *Cytogenetics* **10:** 81.
Aula, P. (1970). *Hereditas* **65:** 163.
Aya, T., and Sandberg, A. A. (1971). *J. Nat. Cancer Inst.* **47:** 961.
Bajer, A. (1959). *Hereditas* **45:** 579.
Becker, F. F., and Green, H. (1960). *Exp. Cell Res.* **19:** 361.
Benyesh-Melnick, M., Stich, H. F., Rapp, F., and Hsu, T. C. (1964). *Proc. Soc. Exp. Biol. Med.* **117:** 546.
Bonner, J., Dahmus, M. E., Fambrough, D., Huang, R. C., Marushige, K., and Tuan, D. Y. H. (1968). *Science* **159:** 47.
Borun, T. W., Scharff, M. D., and Robbins, E. (1967). *Proc. Nat. Acad. Sci. U.S.* **58:** 1977.
Boss, J. (1954). *Exp. Cell Res.* **7:** 443.
Boss, J. (1955). *Exp. Cell Res.* **8:** 181.
Brown, S. W. (1954). *Univ. Calif., Berkeley, Publ. Botany* **27:** 231.
Brown, S. W. (1966). *Science* **151:** 417.
Byvoet, P. (1966). *J. Mol. Biol.* **17:** 311.
Cantell, K., Saksela, E., and Aula, P. (1966). *Ann. Med. Exp. Biol. Fenn.* **44:** 255.
Cantor, K. P., and Hearst, J. E. (1966). *Proc. Nat. Acad. Sci. U.S.* **55:** 642.
Cantor, K. P., and Hearst, J. E. (1970). *J. Mol. Biol.* **49:** 213.
Chevaillier, P. H. (1969). *Exp. Cell Res.* **58:** 213.
Clark, R. J., and Felsenfeld, G. (1971). *Nature (London), New Biol.* **229:** 101.
Cleveland, L. R. (1949). *Trans. Amer. Phil. Soc.* **39:** 1.
Comings, D. E. (1967). *J. Cell Biol.* **35:** 699.
Dallam, R. D., and Thomas, L. E. (1953). *Biochim. Biophys. Acta* **11:** 79.
Das, N. K. (1962). *J. Cell Biol.* **15:** 121.
Davidson, D., and Anderson, N. G. (1960). *Exp. Cell Res.* **20:** 610.
Dingman, C. W., and Sporn, M. B. (1964). *J. Biol. Chem.* **239:** 3483.
DuPraw, E. J. (1968). "Cell and Molecular Biology," p. 514. Academic Press, New York.
DuPraw, E. J. (1970). "DNA and Chromosomes." Holt, New York.
DuPraw, E. J., and Rae, P. M. M. (1966). *Nature (London)* **212:** 598.
Flemming, W. (1880). *J. Cell Biol.* **25**, Suppl.: 3 (1965). [Engl. Transl.]
Gay, H. (1956). *J. Biophys. Biochem. Cytol., Suppl.* **2:** 407.
Glaser, R., and Gabbay, E. J. (1968). *Biopolymers* **6:** 243.
Gurdon, J. B. (1968). *J. Embryol. Exp. Morphol.* **20:** 401.
Gurley, L. R., Hardin, J. M., and Walters, R. A. (1970). *Biochem. Biophys. Res. Commun.* **38:** 290.
Hagen, U. (1960). *Nature (London)* **187:** 1123.
Hancock, R. (1969). *J. Mol. Biol.* **40:** 457.
Harnden, D. G. (1964). *Amer. J. Hum. Genet.* **16:** 204.
Hartmann, M. (1928). *Zool. Jahrb., Abt. Allgem. Zool. Physiol. Tiere* **45:** 973.
Heilbrunn, L. V. (1952). *In* "Modern Trends in Physiology and Biochemistry" (E. S. G. Baron, ed.), p. 123. Academic Press, New York.
Hertwig, R. (1908). *Arch. Zellforsch.* **1:** 1.
Howard, A., and Pelc, S. R. (1953). *Heredity* **6:** 261.

Huberman, J. A., and Attardi, G. (1966). *J. Cell Biol.* **31**: 95.
Ikeuchi, T., and Sandberg, A. A. (1970). *J. Nat. Cancer Inst.* **45**: 951.
Ikeuchi, T., Sanbe, M., Weinfeld, H., and Sandberg, A. (1971). *J. Cell Biol.* **51**: 104.
Jacobson, W., and Webb, M. (1952). *Exp. Cell Res.* **3**: 163.
Johnson, R. T., and Rao, P. N. (1970). *Nature (London)* **226**: 717.
Johnson, R. T., and Rao, P. N. (1972). *Advan. Biosci.* **8**: 237.
Johnson, R. T., Rao, P. N., and Hughes, S. D. (1970). *J. Cell Physiol.* **76**: 151.
Kao, F. T., and Puck, T. T. (1970). *Nature (London)* **228**: 329.
Kato, H., and Sandberg, A. A. (1967). *J. Cell Biol.* **34**: 35.
Kato, H., and Sandberg, A. A. (1968a). *J. Nat. Cancer Inst.* **40**: 165.
Kato, H., and Sandberg, A. A. (1968b). *J. Nat. Cancer Inst.* **41**: 1117.
Kato, H., and Sandberg, A. A. (1968c). *J. Nat. Cancer Inst.* **41**, 1125.
Kaufmann, B. P., McDonald, M., and Gay, H. (1948). *Nature (London)* **162**: 814.
Lake, R. S., Goidl, J. A., and Salzman, N. P. (1972). *Exp. Cell Res.* **73**: 113.
Lima-de-Faria, A. (1959). *J. Biophys. Biochem. Cytol.* **6**: 457.
Liquori, A. M., Constantino, L., Crescenzi, V., Elia, V., Giglio, E., Poluti, R., de Santis Savino, M., and Vitagliano, V. (1967). *J. Mol. Biol.* **24**: 113.
Love, R. (1957). *Nature (London)* **180**: 1338.
Lyon, M. F. (1962). *Amer. J. Hum. Genet.* **14**: 135.
Lyon, M. F. (1972). *Biol. Rev. Cambridge Phil. Soc.* **47**: 1.
Maio, J. J., and Schildkraut, C. L. (1967). *J. Mol. Biol.* **24**: 29.
Matsui, S., Weinfeld, H., and Sandberg, A. A. (1971). *J. Nat. Cancer Inst.* **47**: 401.
Matsui, S., Yoshida, H., Weinfeld, H., and Sandberg, A. A. (1972). *J. Cell Biol.* **54**: 120.
Mazia, D. (1956). *Amer. Sci.* **44**: 1.
Mazia, D. (1963). *J. Cell. Comp. Physiol.* **62**, Suppl. 1: 123.
Miles, C. P., and O'Neill, F. (1969). *J. Cell Biol.* **40**: 533.
Mitchison, J. M. (1971). "The Biology of the Cell Cycle." Cambridge Univ. Press, London and New York.
Mittwoch, U. (1967). *Cytogenetics* **6**: 38.
Nagl, W. (1970). *Caryologia* **23**: 71.
Nevo, A. C., Mazia, D., and Harris, P. J. (1970). *Exp. Cell Res.* **62**: 173.
Neyfakh, A. A., Abramova, N. B., and Bagrova, A. M. (1971). *Exp. Cell Res.* **65**: 345.
Nichols, W. W., Levan, A., and Aula, P. (1964). *Hereditas* **51**: 380.
Nichols, W. W., Levan, A., and Aula, P. (1965). *Hereditas* **54**: 101.
Nichols, W. W., Aula, P., Levan, A., Heneen, W., and Norrby, E. (1967). *J. Cell Biol.* **35**: 257.
Ohlenbusch, H. H., Olivera, B. M., Tuan, D., and Davidson, N. (1967). *J. Mol. Biol.* **25**: 299.
Ohno, S. (1969). *Annu. Rev. Genet.* **3**: 495.
O'Neill, F., and Miles, C. P. (1970). *Proc. Soc. Exp. Biol. Med.* **134**: 825.
Östergren, G. (1961). *Chromosoma* **12**: 80.
Östergren, G., and Wakonig, T. (1954). *Bot. Notis.* **4**: 357.
Östergren, G., and Bajer, A. (1961). *Chromosoma* **12**: 72.
Patil, S., Rao, P. N., and Lubs, H. (1972). *Mammalian Chromosome Newslett.* **13**: 91.

Pederson, T. (1972). *Proc. Nat. Acad. Sci. U. S.* **69**: 2224.
Pederson, T., and Robbins, E. (1970). *J. Cell Biol.* **47**: 155a.
Pederson, T., and Robbins, E. (1972). *J. Cell Biol.* **55**: 322.
Philpot, J. St. L., and Stanier, J. E. (1957). *Nature (London)* **179**: 102.
Prescott, D. M. (1955). *Exp. Cell Res.* **9**: 328.
Prescott, D. M. (1956). *Exp. Cell Res.* **11**: 94.
Prescott, D. M., and Bender, M. A. (1963). *J. Cell. Comp. Physiol.* **62**, Suppl. 1: 175.
Prescott, D. M., and Goldstein, L. (1968). *J. Cell Biol.* **39**: 404.
Rao, P. N. (1969). *Exp. Cell Res.* **57**: 230.
Rao, P. N., and Johnson, R. T. (1970). *Nature (London)* **225**: 159.
Rao, P. N., and Johnson, R. T. (1971). *J. Cell. Physiol.* **78**: 217.
Rao, P. N., and Johnson, R. T. (1972a). *J. Cell Sci.* **10**: 495.
Rao, P. N., and Johnson, R. T. (1972b). *J. Cell Biol.* **55**: 212a.
Rao, P. N., and Johnson, R. T. (1974). In preparation.
Richards, B. M., and Bajer, A. (1961). *Exp. Cell Res.* **22**: 503.
Ris, H. (1956). *J. Biophys. Biochem. Cytol., Suppl.* **2**: 385.
Robbins, E., and Borun, T. W. (1967). *Proc. Nat. Acad. Sci. U.S.* **57**: 409.
Robbins, E., and Gonatas, N. K. (1964). *J. Cell Biol.* **21**: 429.
Robbins, E., and Pederson, T. (1970). *J. Cell Biol.* **47**: 172a.
Robbins, E., and Scharff, M. D. (1966). *In* "Cell Synchrony: Studies in Biosynthetic Regulation" (I. L. Cameron and G. M. Padilla, eds.), p. 353. Academic Press, New York.
Robbins, E., Pederson, T., and Klein, P. (1970). *J. Cell Biol.* **44**: 400.
Roizmann, B. (1962). *Cold Spring Harbor Symp. Quant. Biol.* **27**: 327.
Sadgopal, A., and Bonner, J. (1970). *Biochim. Biophys. Acta* **207**: 227.
St. Amand, G. A., Anderson, N. G., and Gaulden, M. E. (1960). *Exp. Cell Res.* **20**: 71.
Salzman, N. P., Moore, D. E., and Mendelsohn, J. (1966). *Proc. Nat. Acad. Sci. U.S.* **56**: 1449.
Sanbe, M., Aya, T., Ikeuchi, T., and Sandberg, A. A. (1970). *J. Nat. Cancer Inst.* **44**: 1079.
Sandberg, A. A., Sofuni, T., Takagi, N., and Moore, G. E. (1966). *Proc. Nat. Acad. Sci. U.S.* **56**: 105.
Sandberg, A. A., Aya, T., Ikeuchi, T., and Weinfeld, H. (1970). *J. Nat. Cancer Inst.* **45**: 615.
Scaife, J. F., and Brokee, H. (1969). *Can. J. Biochem.* **47**: 237.
Scherbaum, O., and Zeuthen, E. (1954). *Exp. Cell Res.* **6**: 221.
Schwartz, A. G., Cook, P. R., and Harris, H. (1971). *Nature (London), New Biol.* **230**: 5.
Spalding, J., Kajiwara, K., and Mueller, G. C. (1966). *Proc. Nat. Acad. Sci. U.S.* **56**: 1535.
Steffensen, D. M. (1961). *Int. Rev. Cytol.* **12**: 163.
Stein, G., and Baserga, R. (1970). *Biochem. Biophys. Res. Commun.* **41**: 715.
Stenman, S. (1971). *Exp. Cell Res.* **69**: 372.
Stenman, S., and Saksela, E. (1969). *Hereditas* **62**: 323.
Stenman, S., and Saksela, E. (1971). *Hereditas* **69**: 1.
Stubblefield, E. (1964). *In* "Cytogenetics of Cells in Culture" (R. J. C. Harris, ed.), Symposia of the International Society for Cell Biology, Vol. 3, p. 223. Academic Press, New York.

Stubblefield, E. (1966). *J. Nat. Cancer Inst.* **37**: 799.
Stubblefield, E. (1973). *Int. Rev. Cytol.* **35**: 1.
Suwalsky, M., Traub, W., Schmneli, U., and Subirana, J. A. (1969). *J. Mol. Biol.* **42**: 363.
Swann, M. M. (1957). *Cancer Res.* **17**: 727.
Swift, H. (1950). *Proc. Nat. Acad. Sci. U.S.* **36**: 643.
Takagi, N., Aya, T., Kato, H., and Sandberg, A. A. (1969). *J. Nat. Cancer Inst.* **43**: 335.
Unakul, W., Johnson, R. T., Rao, P. N., and Hsu, T. C. (1973). *Nature (London), New Biol.* **242**: 106.
Waterson, A. P. (1962). *Nature (London)* **193**: 1163.
Whitfield, J. F., and Youdale, T. (1966). *Exp. Cell Res.* **43**: 602.
Whitfield, J. F., Rixon, R. H., and Youdale, T. (1962). *Exp. Cell Res.* **27**: 143.
Wolff, S. (1969). *Exp. Cell Res.* **57**: 457.
Zakharov, A. F., and Egolina, N. A. (1972). *Chromosoma* **38**: 341.
Zur Hausen, H. (1967). *J. Nat. Cancer Inst.* **38**: 683.

STRUCTURAL "BANDS" IN HUMAN CHROMOSOMES*

G. F. Bahr and
P. M. Larsen

ARMED FORCES INSTITUTE OF PATHOLOGY
WASHINGTON, D.C.

I. Introduction

In a recent article by Bahr *et al.* (1973) possible steps in the condensation of human chromatin to chromosomes have been illustrated with transmission electron micrographs of surface-spread and critical point-dried chromatin from peripheral lymphocytes. It was noted that chromosome fibers may arbitrarily be considered in two groups: longitudinal fibers, and chromomeric fibers. Both are a part of the single long chromatin fiber from which a chromatid arises by folding (DuPraw, 1970).

Chromomeres† are accumulations of fiber loops. A chromomere may occupy the full width of the chromatid or may be smaller. It may be

* The opinions or assertions contained herein are the private views of the authors and are not to be construed as official or as reflecting the views of the Department of the Army or the Department of Defense.

† "Chromomeres: Areas of different optical density and/or different diameters along the length of a chromosome, especially clearly discernible in prophase of meiosis" (McKusick, 1969).

located coaxially with the imaginary chromatid axis or off-axis. Chromomeres are connected by longitudinal parts of the chromatin fiber. About 8–15 longitudinal fibers can be counted or measured at suitable places by scanning densitometry (DuPraw and Bahr, 1969; Bahr, 1973).

Because one chromatid is a strict mirror image of the other, both structurally and molecularly, the position and size of chromomeres along the chromatid axes in a chromosome are also identical and symmetric around the chromosomal plane of symmetry (Bahr, 1974).

Moreover, since quantity of object mass penetrated by the electron beam and contrast are proportional, large chromomeres produce more contrast. As a result, the natural contrast in the discontinuous distribution of chromatin along a chromosome produces the impression of banding.

The relation of this structural banding of chromosomes to the banding observable in a fluorescent microscope after quinacrine mustard staining (Caspersson *et al.*, 1970a,b; Zech, 1973), and to the Giemsa-based banding techniques (e.g., Hsu *et al.*, 1972; Schnedl, 1973), has been the subject of the earlier mentioned work by Bahr *et al.* (1973). At the time of that study too few electron micrographs of chromosomes had been analyzed to warrant rendering a definitive account. It was nevertheless possible to state that the banding of chromosomes, as it is seen by light microscopists, is based on the distribution of both chemical *and* structural properties along chromosomes.

Since then some 500 electron micrographs of chromosomes have been selected from among more than 5000 and subjected to identification procedures. The chromosomes were taken from eight different subjects. A more definitive analysis of structural banding in human chromosomes can therefore be presented in this chapter.

II. Materials and Methods

Peripheral human lymphocytes were cultured according to standard techniques using Difco reagents and media. Colchicine was added 5 hours before harvesting. The cells were hypotonically treated and centrifuged, and a pellet of swollen cells was released onto the surface of distilled water. Material was then picked up at random by touching Formvar or carbon-coated grids for electron microscopy, to the surface. The grids so charged with nuclear material were immediately immersed in 30% ethanol, dehydrated in graded ethanol, and critical-point dried. Details of this technical step have recently been described (Bahr, 1973).

Electron micrographs were taken at 2100 times magnification and constant exposure. They were processed by standard photochemical procedures (Bahr, 1973). An Ortholux* microscope was fitted with an MPV-1 optical chain,* a 1× objective, and a 1× condenser. The tungsten light source was powered by a highly stabilized supply. A 1P28 photomultiplier with a photovolt† power supply rendered the current output for driving a small strip chart recorder on one axis.

In the MPV, a rectangular measuring diaphragm was adjusted in width to 0.05 μm in the original object, i.e., to the width of an average minor band. The length of this slit was adjusted for each micrograph to cover the entire cross section of a chromatid. It was moved over the length of the chromatid by a constant-speed motor attached to the stage drive,* and guided by hand whenever the chromatid was bent. In a few instances, sister chromatids were so close to each other that a clear separation of the two in scanning was impossible. Scans were nevertheless obtained according to best judgment.

Amplification, both to and in the strip-chart recorder, was adjusted to render full pen excursion per scan. However, since the cross section of chromatids varies with the degree of chromosomal contraction, the areas under all scans are only roughly comparable with respect to the total mass they represent. Each scan is consistent in itself, since it is obtained with a fixed set of working parameters, and therefore the mass of any band is inherently represented in true proportion to other bands. Equally comparable are the masses of short and of long arm, as well as that of the whole chromosome.

A short scan through the electrographic density in the object's background served to determine background transmission levels at the beginning and end of each chromatid. Background transmission is used to set the lower limit on the excursions of the recording pen, while maximum transmission in the scan is set to almost full pen excursion on the recorder by adjusting amplification of photocurrent. Background transmission renders the baseline for each curve. Markers indicate intersections of the curve ends with the baseline, and the position of the centromere.

III. Results

Those working with banding techniques at the light microscopic level have repeatedly stated that optimal banding patterns are obtained from

* E. Leitz, Inc., Rockleigh, New Jersey.
† Photovolt Corporation, New York.

chromosomes which are half-way between late prophase and metaphase (e.g., Caspersson *et al.,* 1971). At full condensation narrow bands are fused to larger blocks of bands. During early prophase, chromosomes are very long and slender, often bent or curving, with hardly recognizable chromatids. The same holds true for chromosomes under the electron microscope (Fig. 10, Bahr *et al.,* 1973). For this study a set of chromosomes of medium condensation were selected. It is believed that these are comparable to the banded chromosomes of light microscopy. Another set of fully condensed chromosomes was also selected for scanning. This

TABLE I
MEASUREMENTS AND COUNTS FROM ELECTRON MICROGRAPHS OF CRITICAL-POINT DRIED HUMAN CHROMOSOMES

No.	Relative length[a]	Centromere index (CI)[b]	Coefficient of variation of CI (%)	Cp[c]	Cq[c]	P+q	Chromomere index[d]
A1	364/8.43	47.27	9.56	14	14	28	13.0
A2	338/7.83	39.90	7.71	9	14	23	14.7
A3	279/6.46	46.61	12.33	10	11	21	13.3
B4	260/6.02	31.56	9.84	5	11	16	16.3
B5	248/5.75	30.55	6.69	4	12	16	15.5
C6	243/5.63	41.90	3.79	6	10	16	15.2
C7	217/5.03	39.50	8.07	5	8	13	16.7
C8	198/4.59	31.16	5.70	4	9	13	15.2
C9	191/4.43	35.79	6.71	4	8	12	15.9
C10	192/4.45	32.77	7.66	4	8	12	16.0
C11	193/4.47	40.42	6.69	5	7	12	16.1
C12	180/4.17	30.54	10.88	3	8	11	16.3
D13	155/3.59	14.48	19.78	2	8	10	15.5
D14	150/3.48	15.73	12.00	2	7	9	16.7
D15	132/3.06	13.94	14.14	1	7	8	16.5
E16	125/2.90	40.06	7.79	3	6	9	13.9
E17	119/2.76	33.62	6.47	3	5	8	14.9
E18	111/2.57	30.10	7.56	2	5	7	15.9
F19	98/2.27	45.10	7.19	2	3	5	19.6
F20	90/2.09	44.20	5.64	3	3	6	15.0
G21	68/1.58	30.05	12.70	1	3	4	17.0
G22	72/1.67	31.97	11.08	2	4	6	12.0
X	205/4.75	42.06	6.58	6	8	14	14.6
Y	88/2.04	19.81	17.22	1	4	5	17.6

[a] In arbitrary units/and in percent of total length.
[b] Length of short arm divided into total length (satellites not included).
[c] Counts of chromomeres in short (p) and long (q) arm.
[d] Length of chromosome divided by the number of chromomeres.

TABLE II
COEFFICIENTS OF VARIATION (CV) FOR CENTROMERE INDEX MEASURED BY LIGHT AND ELECTRON MICROSCOPY

No.	Light microscopy[a] CV	Electron microscopy CV
A1	5.34	9.56
A2	6.68	7.71
A3	4.43	12.33
B4	11.87	9.84
B5	9.70	6.69
C6	6.59	3.79
C7	11.35	8.07
C8	8.53	5.70
C9	12.53	6.71
C10	8.97	7.66
C11	8.14	6.69
C12	14.59	10.88
D13	21.68	19.78
D14	21.19	12.00
D15	26.13	14.14
E16	13.17	7.79
E17	10.34	6.47
E18	15.78	7.56
F19	8.90	7.19
F20	5.48	5.64
G21	17.48	12.70
G22	23.04	11.08
X	7.20	6.58
Y	22.07	17.22

[a] From Caspersson and Zech (in Hamerton *et al.*, 1973).

set is, in many respects, comparable to classical chromosomes of the prebanding times of cytogenetics. The nature of sampling for whole-mounting surface-spread chromosomes entails collecting mostly groups of 5–15 or single chromosomes. Whole metaphase plates are seldom collected because their image is often partially masked by the grid bars of the specimen carrier.

Consequently, this study was undertaken with the objective of identifying all members of the human karyotype from electron micrographs of single chromosomes. The length of the short arm, divided into the chromosome length, was used as one identifying feature (Centromere Index). A comparison of the coefficient of variation of centromeric indices from this study (Tables I and II), with those from Table 5 of the "Standardization in Human Genetics," Paris Conference, 1971 (Hamerton *et al.*, 1973), as well as a comparison of relative chromosomal

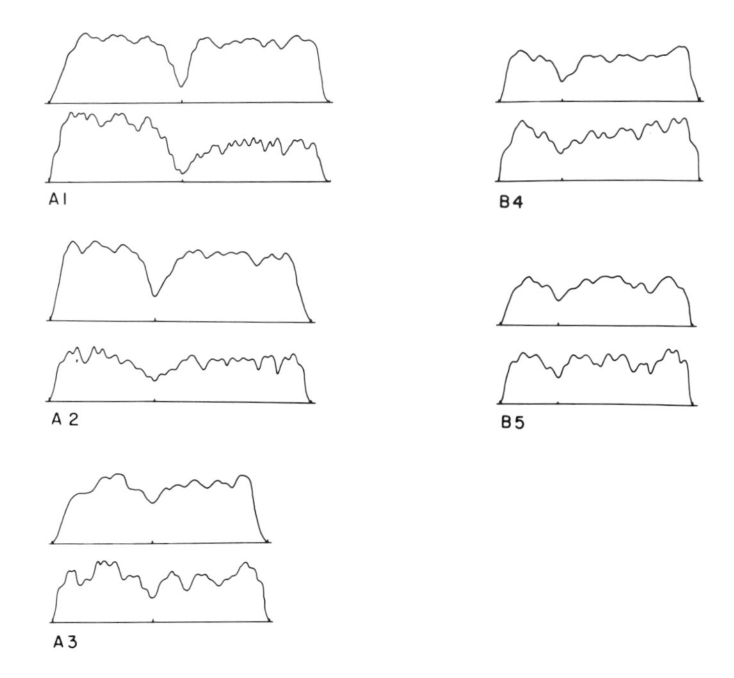

FIG. 1. See caption on facing page.

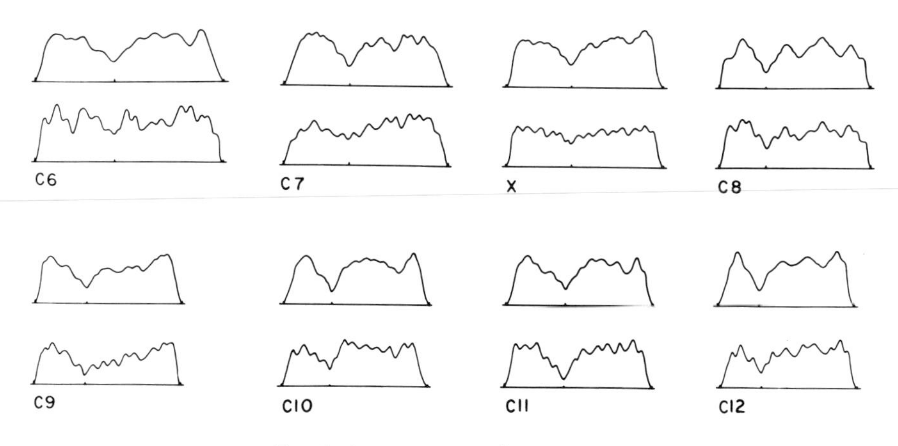

FIG. 2. See caption on facing page.

lengths (in percent), suggests that a good classification of chromosomes has been accomplished. It can be seen (Table II) that, on the average, electron microscopy presents a lower variation of centromere index than light microscopy.

Nine to ten single chromosomes were selected for each of the condensed

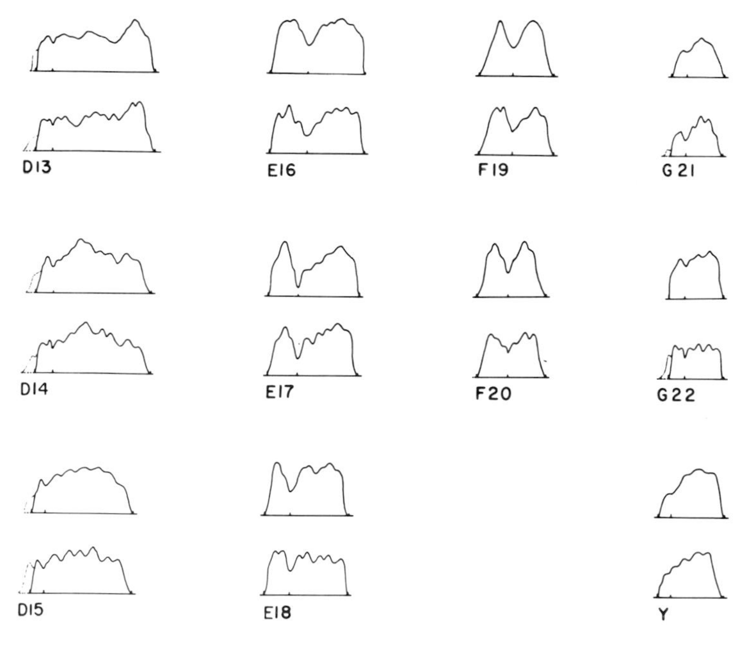

FIG. 3.

FIGS. 1–3. Summary presentation of transmission scans along the chromatids of 9 to 10 chromosomes from electron micrographs of whole-mounted specimens. Upper curves represent normally condensed chromosomes; lower curves, prophasic extended chromosomes. The length of upper and lower curves has been normalized to the average length of the condensed state. Chromosomes were not stained. The relative height along each curve reflects the distribution of dry matter (chromatin) along an average chromatid. Satellites are indicated by broken lines. AFIP Neg 74-6332-1, 74-6332-2, 74-6332-3.

and extended states of a specific chromosome. For reasons to be discussed later, electron microscopic transmission scans do not lend themselves easily to automated analysis. Therefore transmission scans of a chromosome were matched by superimposing the tracings of sister chromatids and of chromatids from all in the group. Visual judgment dominated the assessment of similarities (Goldmeier, 1972). At the end, composite curves were drawn and once again compared to the originals to ensure that they were representative.

In the course of this identification process the published human karyotypes of Q-, G-, and R-banded chromosomes were initially consulted. It was soon found that chromomere patterns under the electron microscope do not provide a close match with any published light microscopic pattern, but have many common traits with each. It was therefore decided to use only the measurements of centromere index and relative length as guides and proceed on an independent basis.

Figures 1–3 illustrate the average features of transmission scans for each chromosome of the human karyotype. The upper tracings are from condensed chromosomes, the lower ones from extended ones. Both are presented in proper length relative to the entire karyotype. Curves from extended chromosomes have been reduced to the respective length of the condensed form, in order to permit a comparison of features in both states of condensation.

First, it will be noted that none of the scans show gaps, interband regions, or negative bands. This is due to the fact that a chromatid is a solid body, and that scans from electron micrographs represent only dry mass modulations of this body in three dimensions. Electron micrographs represent only the dry mass transsected by the electron beam. The recorded curves of Figs. 1–3 are all reflections of the distribution of chromatin mass along a chromatid. Only at the centromere does the mass diminish markedly in cross section, as expected from the well-known shape of a chromosome at metaphase.

Chromomeres, then, are not as pronounced as typical bands may be under the light microscope. It should also be kept in mind that bands stained with a light-absorbing stain, such as Giemsa or Feulgen, increase and decrease contrast *exponentially,* with variation in mass (Lambert-Beer's law), while transmission in electron micrographs is *linearly* related to dry mass.

Every maximum and every shoulder in the curves of Figs. 1–3 represents a chromomere reproducibly observed in the set of curves for each chromosome. For reasons to be discussed later a chromomere is rarely seen at precisely the same location in two chromosomes. The positions are comparable and identifiable, however, and represent averages of observations indicated in the curves.

After some study of Figs. 1–3, the unique features of chromomere structure (banding) associated with most chromosomes can be detected. Features of arm length and "banding" serve to identify all chromosomes in the karyotype. The lack of full metaphases at this time precludes comparison of homologs.

The following is a detailed description of each chromosome. Groups of chromomeres are called complexes in extended chromosomes, and blocks in condensed ones. Numbers of chromomeres for short and long arms are indicated by Cp and Cq, respectively.

A1. The short arm of the extended state indicates the grouping of chromomeres into three complexes. The most distal complex consists of at least 5 chromomeres, while the central complex comprises four chromomeres. The most proximal complex has five chromomeres of decreasing size toward a rather pronounced centromere.

In the long arm four complexes are seen. The most distal is composed of three prominent chromomeres. A marked interchromomeric region separates it from the more distal complex, which consists of only two chromomeres. Most centrally in the long arm a complex of 5 relatively small and evenly spaced chromomeres is located. The complex nearest the centromere consists of four chromomeres decreasing in size toward the centromere.

The condensed state for the short arm exhibits three modulated blocks corresponding to the complexes seen in the extended state. Also for the long arm a corresponding number of four blocks can be discerned. The centromere remains prominent. Cp = 14, Cq = 14.

A2. The short arm of the extended form consists of three complexes of three chromomeres each, distinctly separated by interchromomeric regions. The distal chromomeres are large but gradually decrease in size toward the centromere. There is a comparable increase in chromomeric size in the proximal complex of the long arm involving 4 chromomeres. The proximal middle complex incorporates 5 chromomeres with a marked partition between the first 2 and last 3 chromomeres. The next to distal complex consists of two chromomeres of the same size, the distal one of three. Interchromomeric regions between the distal three complexes are always marked and aid in the identification of the A2 chromosome.

In the condensed state only blocks of chromatin with some remaining indication of chromomeric structure are discernible. Most prominent interchromomeric regions appear now between. the two proximal and the two distal blocks of the long arm. Cp = 9, Cq = 14.

A3. There are three chromomeric complexes in each chromosome arm. The most distal in the short arm consists of 3 chromomeres, the middle of 4, and the proximal of 3. The middle complex is the most massive of the three. The centromere is weakly expressed. Three, three, and five chromomeric complexes (3, 3, 5) follow distally in the long arm, of which the distal one is the largest. They are well separated with increasing interchromomeric regions toward the centromere.

In the condensed state complexes are fused to blocks which remain well distinguishable in the long arm but are difficult to discern in the short arm. Cp = 10, Cq = 11.

B4. In the extended form of this chromosome 5 chromomeres arranged in two complexes constitute the short arm, which is smaller toward the centromere. Ten chromomeres occupy the long arm, which are larger the more distally they are located. This is a characteristic feature of this chromosome. The chromomeres of the long arm are grouped in four complexes with four, two, two (4, 2, 2), and three chromomeres from centromere to telomere.

The condensed state reflects the two complexes of the short arm as two chromatin blocks. In the long arm, three rather than four blocks can be discerned because the two most distal ones have merged into one block. Cp = 5, Cq = 11.

B5. The short arm of this chromosome consists of four chromomeres constituting one single complex. In the long arm, three complexes and one double chromomere are distinguishable. The most proximal complex consists of four chromomeres, separated from the next complex by a moderately wide interchromomeric region. The next distal complex consists of two prominent chromomeres. It is followed by a rather small, double chromomere, well separated from its neighbors. The most distal complex in the long arm is composed of four chromomeres; it is the most prominent complex in the long arm.

The condensed state mirrors some structural detail of the short arm. In the long arm the two proximal complexes and the double chromomere have fused to render one block. A distinct interchromomeric region separates the distal chromatin block from the rest of the long arm. Cp = 4, Cq = 12.

C6. There are two complexes of three chromomeres each in the short arm. The proximal set of chromomeres decreases in size toward the centromere. A prominent complex of the chromomeres follows proximally in the long arm. Thereafter a rather small and a prominent complex of two chromomeres each are located distally. The most total complex in the long arm consists of three chromomeres. The centromere is not distinguished.

The condensed state shows fusion of all chromomeres in the short and long arms; only the distal end of the long arm remains as a separate block. Cp = 6, Cq = 10.

C7. The extended state shows five chromomeres in the short arm without noticeable complex formation. The long arm consists of three complexes. The most proximal and the second feature two chromomeres each of increasing size. In the distal complex four chromomeres are located.

When the chromosome condenses, only one block is seen in the short arm and two blocks in the long arm. A distinct interchromomeric region remains between the blocks of the long arm. Cp = 5, Cq = 8.

C8. Extended: One complex of four chromomeres in the short arm, three with three complexes each in the long arm.

This distinctive partition into complexes appears as blocks in the condensed state. A small chromomere at the distal end of the short arm adds to the recognition features. Cp – 4, Cq – 9.

C9. The extended short arm consists of one complex of four chromomeres, the largest one being in the next to distal position. Chromomeres are smaller toward the centromere, giving the chromosome a characteristic appearance. The eight chromomeres of the long arm also show a characteristic size increase toward the distal end. They are divided in three complexes of three, two, and three chromomeres.

In the condensed state, mass around the centromere increases slightly. The characteristic distribution, with relatively low mass at the center is maintained. Cp = 4, Cq = 8.

C10. The short arm of the extended chromosome consists of one complex of four chromomeres, quite similar in arrangement to C9. In the long arm two complexes are conspicuous; the proximal consists of six chromomeres with one large chromomere close to the usually small centromeric chromomeres. In the distal complex two chromomeres are discernible. The centromere is prominent.

In the condensed state a marked centromere and a prominent distal block in the long arm serve to identify this chromosome. Cp – 4, Cq = 8.

C11. One complex of five chromomeres is found in the extended short arm. In the long arm two complexes of five and two chromomeres, each of rather even size, are located. There is a tendency in the short arm to split into two complexes.

The separation into complexes in the long arm remains after condensation. The short arm also tends to form two blocks. Cp = 5, Cq = 7.

C12. Three chromomeres form the short arm, three complexes the long arm, with three, two, and three chromomeres each. The relatively even size of all chromomeres is a characteristic of this chromosome.

Upon condensation three blocks are seen in the long arm, while the short arm has fused to one block with a shoulder. Cp = 3, Cq = 8.

D13. The short arm of the extended state consists essentially of two distinguishable chromomeres, plus the features of the satellite. The long arm has three major chromomeric complexes. The first, most proximal one consists of two chromomeres, the second of four, and the third of two quite prominent ones. These distal chromomeres are distinguishing features of D13.

In the condensed state one typically finds merger or fusion of these two. In the densitometric trace only a weak shoulder indicates the presence of the most distal chromomere. The centromere is not prominent.

Only the three blocks of chromatin are distinguishable in the long arm. Cp = 2, Cq = 8.

D14. The short arm of the extended state consists of two chromomeres and a rather well distinguishable satellite. The centromere is moderately expressed and is comparable to D13. In the long arm a small chromomere follows the centromere and marks the beginning of the major chromomere complex which is the distinguishing feature of D14. Its greatest mass is most often located centrally in the long arm. It consists of four chromomeres. The mutual associations of these are highly variable. A moderate interchromomeric region separates this complex from the most distal two chromomeres.

In the condensed state one chromomeric block marks the short arm. The satellite is equally condensed, but clearly discernible. All the chromomeres of the proximal and central long arm are fused to a prominent block. The most distal chromomeres remain separated as a block. Cp = 2, Cq = 7.

D15. The short arm in this chromosome consists of only one chromomere and a prominent satellite. Its centromere is somewhat longer than those of D13 and 14. In the long arm, six rather evenly spaced chromomeres of comparable size are located.

In the condensed chromosome only satellite and centromere can be distinguished. Cp = 1, Cq = 6.

E16. Three chromomeres within one complex constitute the short arm. The long arm does not separate into individual complexes, but is a continuous set of five chromomeres of distally increasing size.

In the condensed state only slight indications of chromomeres in the short and long arm remain. Cp = 3, Cq = 5.

E17. The short arm consists of three chromomeres which are not well separated. It is separated from the long arm by a prominent centromere. In the long arm, a single chromomere is situated proximal to the centromere, distally followed by a complex of four chromomeres, the next to last of which is the largest in this complex.

The condensed state reflects the features of the extended state; however, all chromomeres have now fused to single blocks in short and long arm. Cp = 3, Cq = 5.

E18. In the short arm two chromomeres are clearly distinguished. Two complexes are visible in the long arm. The proximal consists of two chromomeres, the distal of three. There is a characteristic decrease in the mass of chromomeres from the end of the short arm to the end of the long arm.

The condensed state indicates weakly the presence of chromomeres, while the separation of blocks in the long arm remains an easily recognizable feature. Cp = 2, Cq = 5.

F19. This chromosome has two chromomeres in its short arm and three chromomeres in the long arm, forming one complex in the short and one in the long arm.

The condensed state brings total fusion of chromomeres; no detail can be seen any longer. Cp = 2, Cq = 3.

F20. There are three chromomeres each in the short and the long arm of this chromosome. Chromomeres proximal to the centromere are located so that the centromere region is very small in the extended form, but prominent in the condensed form. Cp = 3, Cq = 3.

G21. The extended short arm consists of one large chromomere to which a small satellite is attached. Also three chromomeres are distinguishable in the long arm. The centromere is well marked.

In the condensed state only two blocks of chromatin remain; they are the short and long arm, respectively. Cp = 1, Cq = 3.

G22. This chromosome carries an easily noticeable satellite. Two chromomeres make up the short, and four the long arm. The latter are separated in pairs.

In the condensed state some degree of structural differentiation of chromatin, especially in the long arm remains. While the most distal chromomeres in the long arm are equal or larger than the proximal ones in the extended long arm, the distal block is found to be more massive in the condensed state. Cp = 2, Cq = 4.

X. This chromosome is characterized by the even size of its chromomeres and a rather unconspicuous centromere. Two complexes may be identified in the short arm, each consisting of three chromomeres; three complexes occur in the long arm of two, three, and three chromomeres each, in a distal direction.

Only one block remains in the condensed state in the short arm with a weak indication of the constituent two complexes. The long arm consists of three blocks with a tendency for the most distal one to become rather prominent. Cp = 6, Cq = 8.

Y. The centromere of the Y chromosome is almost obliterated by chromomeric crowding throughout the entire chromosome. One, possibly two, chromomeres are located in the short arm. The long arm has four chromomeres in the characteristic fashion of increasing mass distally on the long arm. Cp = 1, Cq = 4.

In Figs. 4–7 a comparison is made of chromomeric patterns in extended chromosomes (EM) with the published quinacrine mustard pattern (Q) (Caspersson *et al.*, 1970a,b), and scans from a Giemsa karyotype (G) (Schnedl, 1971). For reasons mentioned earlier, a scan of the dry mass (EM) renders a much less profiled curve than the scan of

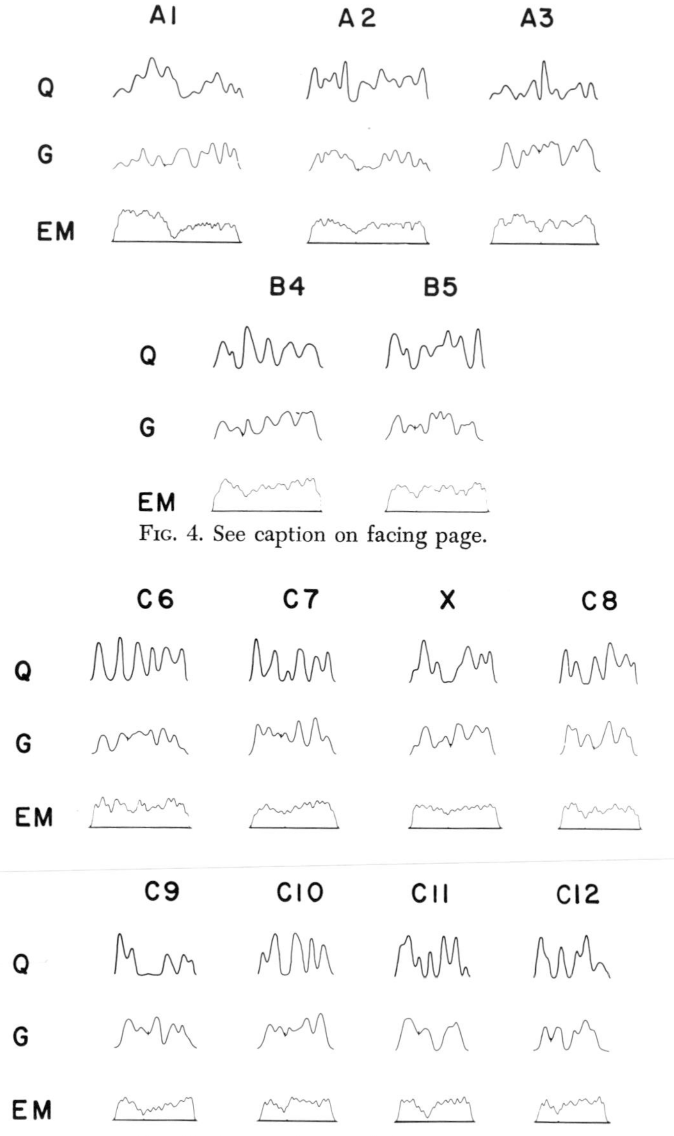

FIG. 4. See caption on facing page.

FIG. 5. See caption on facing page.

a stained chromosome (G). The curves from Q-banded chromosomes have been digitally filtered in Dr. Caspersson's laboratory, and their features thus enhanced.

For many chromosomes a rather good agreement between the location

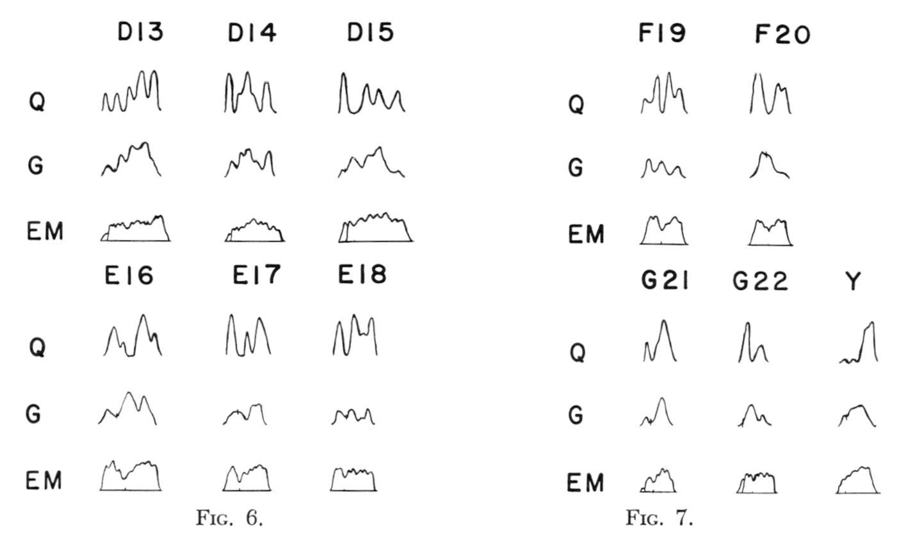

FIG. 6. FIG. 7.

FIGS. 4–7. Comparison of Q = quinacrine mustard stained and fluorescing bands of human chromosomes according to Caspersson and Zech (see text) (these curves are digitally filtered); and of G = densitometric tracings of a Giemsa-banded human karyotype kindly provided by Dr. W. Schnedl (scans were prepared from negative copies analogous to Figs. 1–3); and of the scans of mass distribution in the extended chromosomal state. AFIP Negs. 74-6332-4, 74-6332-5, 74-6332-6, 74-6332-7.

of bands and of chromomeres can be seen. Generally, agreement between Q and EM is better than for G and EM. There are 284 chromomeres illustrated in the EM group. These are considered to be major bands in previously used terminology (Bahr *et al.*, 1973). The Q series in Figs. 4–7 represent 126 bands, and the G series 134. It is therefore understandable if the more detailed EM-curves do not impress the casual viewer with the same clarity as do Q and G bands. Detailed inspection of the three sets of curves, however, will reveal significant relationships essential to understanding the basis of chromosomal banding.

Satellites in groups D and G are represented by broken lines. The size of satellites decreases from D15 to D13. G22 has twice the satellite mass of G21.

IV. Discussion

Human chromosomes (and in our experience the chromosomes from many other species) can be prepared as whole-mounts for electron microscopy. It is difficult to find whole metaphases over a grid-opening. Until preparation techniques have been perfected, only small groups or single chromosomes can be collected. There is no difficulty in accumulat-

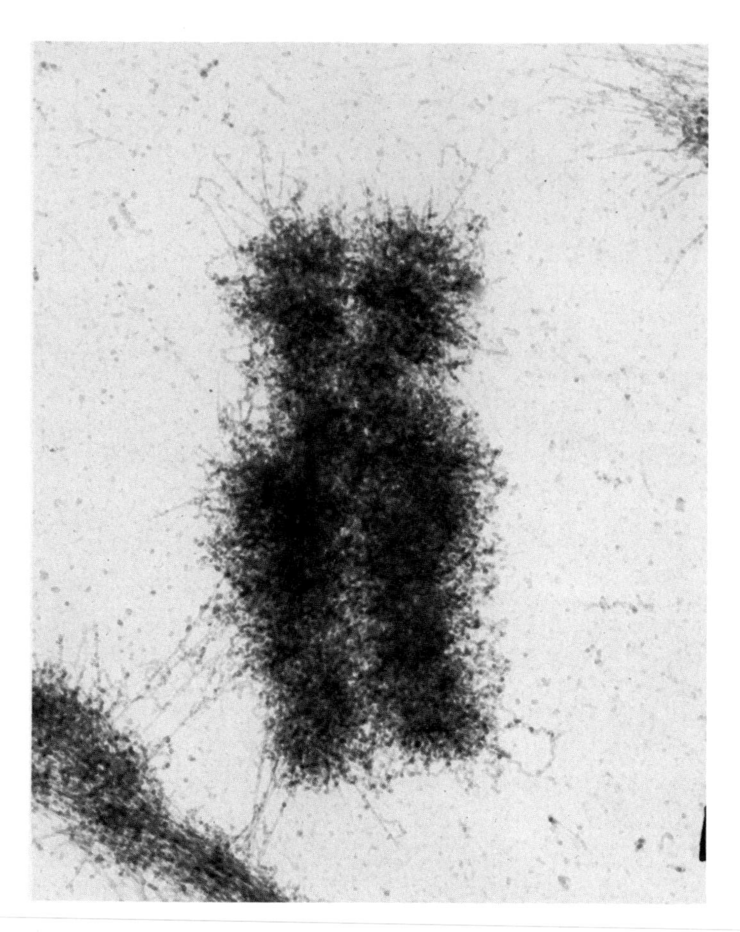

FIG. 8. Normal condensed chromosome No. B5. Structural banding can be appreciated. AFIP Neg. 74-6332-8.

ing literally thousands of single chromosomes for electron microscopy. Among the single chromosomes are many torn and distorted ones, which diminish the yield. Nevertheless, well preserved chromosomes can be obtained in quantity. A relative lack of large chromosomes is noticeable, while small ones are more abundant. This shift in relative frequency is not serious in large series, but may well be with small series. Furthermore, homologs cannot be compared in this manner, nor can aberrations be studied easily. It is thus clear that electron microscopy at the present state of the art cannot be used in routine applications, e.g., on clinical material.

Analysis of chromomere patterns or structural "bands," with the aid of electron microscopy, encounters other difficulties arising from the nature of chromosomal fine structure. Chromomeres are accumulations of fiber loops in the single chromatin fiber constituting one chromatid. The fiber is folded so that it is thrown into loops at one place and continues parallel to the imaginary axis of the chromatid to the next site of looping. In all, this renders a rather loose structure, filling only around 40% of the space of a chromatid. Here and there, touching fibers are cross-linked by suitable protein bonds to provide stability to the structure. Fiber loops emerging and returning to the interior of the chromatid (Fig. 8) are highly mobile.

In well preserved chromosomes, usually of the condensed type, chromomeric regions are *fairly* often found in both chromatids (Figs. 9

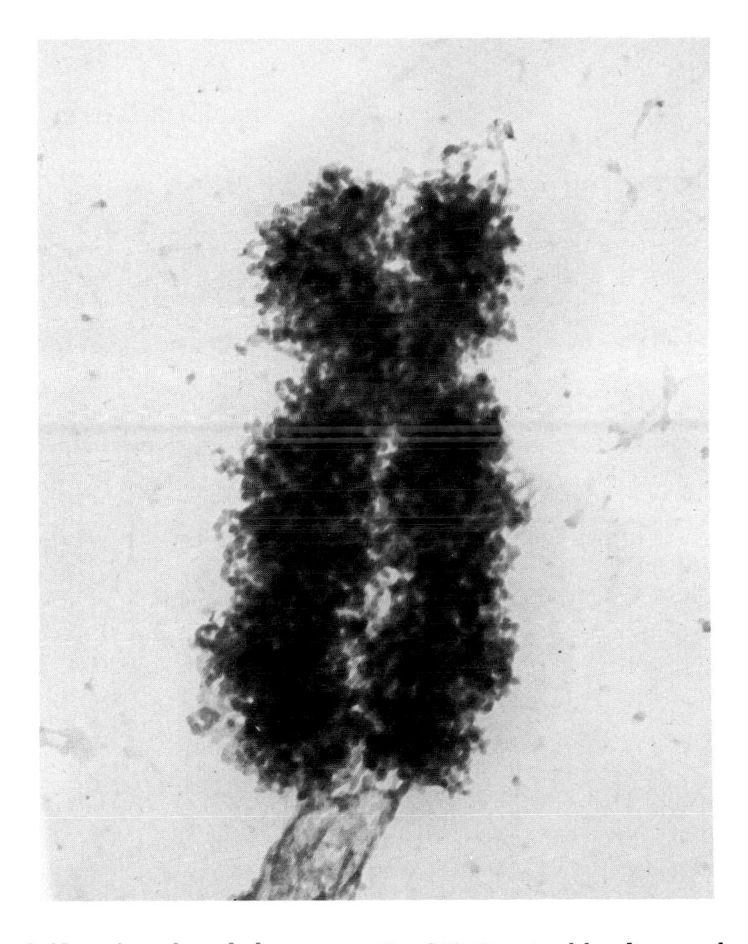

FIG. 9. Normal condensed chromosome No. C12. Structural banding can be appreciated. AFIP Neg. 74-6332-9.

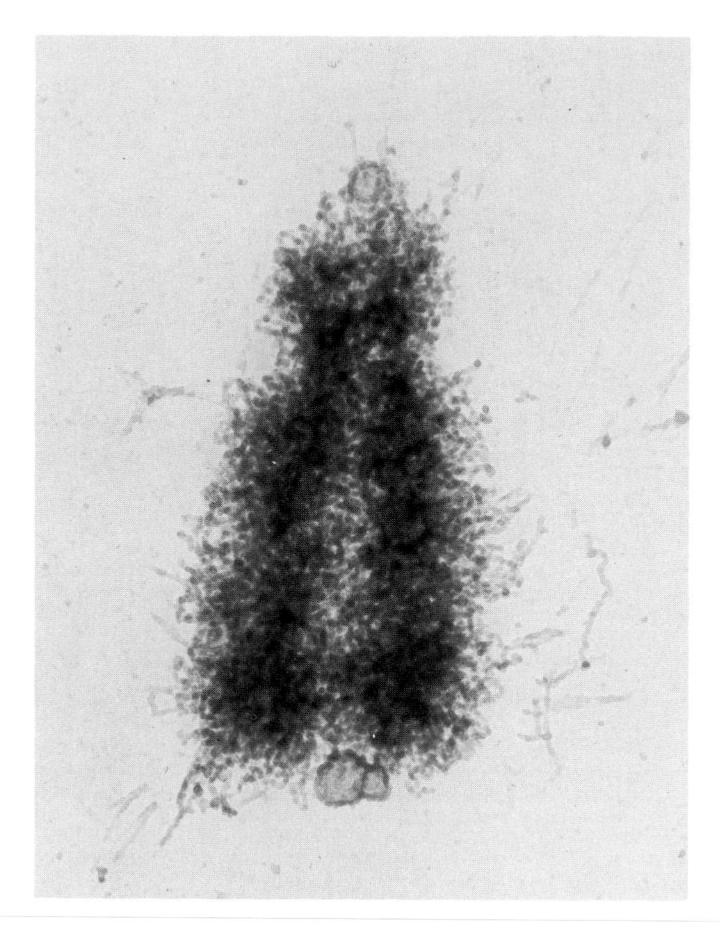

FIG. 10. Normal condensed chromosome No. D13. Note structural banding. AFIP Neg. 74-6332-10.

and 10). More often, the two chromatids are displaced alongside each other, bringing chromomeres out of register. In addition, one chromatid is frequently more extended than the other. An extreme of such differences in chromatids is shown in Fig. 11. Banding in one short arm is revealed, which is largely obscured in the condensed arm. Furthermore, the position of chromomeres is not even around the axis (perpendicular) of a chromatid, but often oblique to this axis or to its side (Fig. 13). The mobility of loops allows for further displacement of chromomeres in the direction of the chromatid axis, resulting in association or fusion with neighboring chromomeres. A chromomere may also be moved off-axis, thus inviting suggestions as to how a chromosome may be organized,

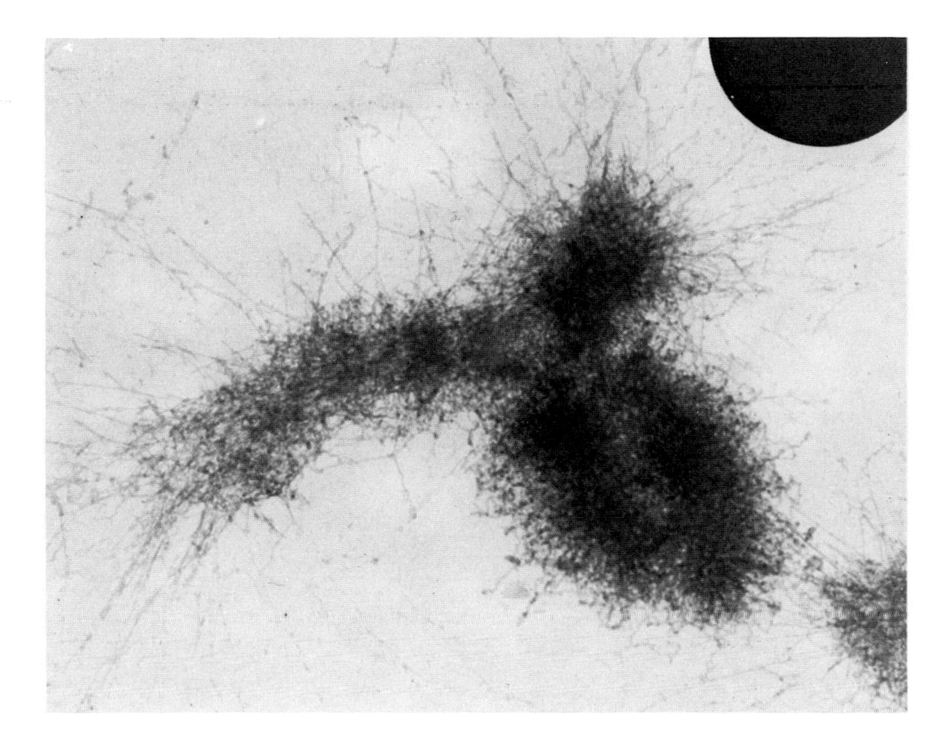

FIG. 11. One short arm of this chromosome has been extended by preparatory forces. Note structural banding. Densities (bands) in the long arm are out of register because of the shift in one long arm. AFIP Neg. 74-6332-11.

the concept of subchromatids (Fig. 12). The shifting of chromomeres or of complexes of chromomeres is the primary reason why computer-assisted pattern recognition routines so far have not produced satisfactory results.

A chromosome appears to have considerable structural integrity, even when isolated and spun in a centrifuge, or in mitosis. It is likely that the single chromatin fiber is not held in its folded shape by conformational forces alone, but also by chemical bonds between touching fibers.

Given these structural considerations, and the fact that the bulk of chromatin consists of protein, it is easy to understand how various treatments and pretreatments are able to influence the appearance and positions of bands under the light microscope. Study of published karyotypes after Giemsa and Feulgen staining shows how variable the position of a band is with respect to the chromatid axis, the same band in the sister chromatid and in the homolog. When a chromosome is treated

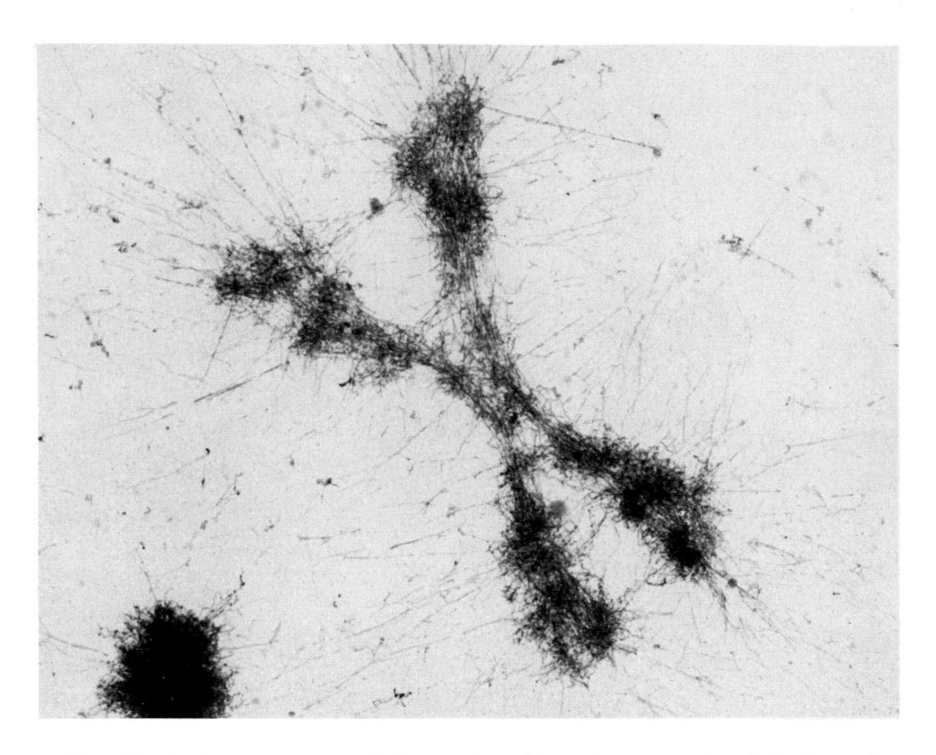

FIG. 12. A chromosome not fully condensed has been spread widely by surface forces of the preparation procedure. Long fiber loops extend from the chromomeric regions. This micrograph is included to illustrate the variable position of chromomeres after the chromosome has been affixed to a surface. Chromomere position varies along chromatid axes and may be off-axis, even giving the impression of subchromatids. AFIP Neg. 74-6332-12.

with reagent mixtures designed to induce banding, one can assume that the delicate and balanced sequence of chromomeres collapses. Conceivably, chromomeres clump in the direction of the strongest bonds with neighboring chromomeres. Those that are connected equally strongly to both sides would associate in a random fashion. Attachment to the substrate (e.g., the object glass) of either one of the two flat sides of a chromosome would add an additional variable.

Our observations shed light on the reasons for well known difficulties in the routine use of Giemsa staining, and most likely any other absorbing stains requiring the rearrangement of chromosomal matter prior to staining. Quinacrine fluorescence microscopy of banded chromosomes has the advantage that no pretreatment is necessary. Thus the structure

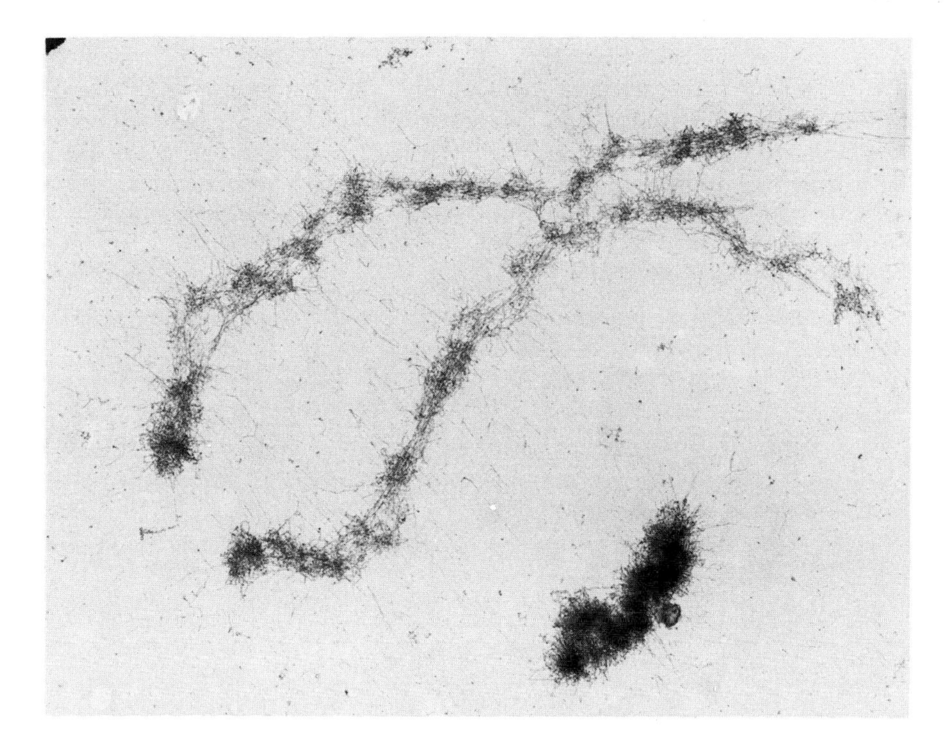

FIG. 13. Maximum extension of a prophasic chromosome. Preparatory forces have moved chromomeres to widely varying positions on and off-axis. AFIP Neg. 74-6332-13

of chromosomes is altered least of all techniques, and therefore renders more consistent results.

V. Conclusion

Chromatids of human chromosomes are structurally differentiated along their length. Accumulations of loops in the single fiber from which a chromatid is folded are called chromomeres. Chromomeres are an expression of variation in the chemical composition of both DNA and protein along the chromatin fiber and thus possess the constancy of a genetic marker. Chromomeres appear as mirroring structures on sister chromatids.

The relatively loose folding of the chromatin fiber allows fiber loops, entire chromomeres, and even complexes of chromomeres to variably associate with neighboring chromomeres as well as to move limited distances in the direction of the imaginary chromatid axis. This relative mobility, taken together with structural associations among chromomeres, allows broad clumps of chromatin to be produced at varying positions along a chromatid. This depends on what treatment or pretreatment has been used to induce banding for light microscopy. The delicate structure of a chromosome can be induced to collapse in various relatively predictable ways.

Without any alteration of chromosomal structure, one can use the position, associations, and size of chromomeres in identifying single chromosomes under the electron microscope. Contrast of chromomeres in electron micrographs produces the impression of banding. It is structural banding.

The bands of chromosomes observed with the light microscope after various treatments appear to have both a structural and a chemical basis. The structural basis of banding implies also that DNA is distributed in the same fashion as has been demonstrated for the dry mass of (DNA containing) chromatin fibers.

REFERENCES

Bahr, G. F. (1973). *In* "Micromethods in Molecular Biology" (V. Neuhoff, ed.), p. 257–284. Springer-Verlag, Berlin and New York.

Bahr, G. F. (1974). *Fed. Proc., Fed. Amer. Soc. Exp. Biol.* (in press).

Bahr, G. F., Mikel, U., and Engler, W. F. (1973). *In* "Chromosome Identification" (T. Caspersson and L. Zech, ed.), p. 280–289. Academic Press, New York.

Caspersson, T., Zech, L., and Johansson, C. (1970a). *Exp. Cell Res.* **62**: 490.

Caspersson, T., Zech, L., Johansson, C., and Modest, E. J. (1970b). *Chromosoma* **30**: 215.

Caspersson, T., Lomakka, G., and Zech, L. (1971). *Hereditas* **67**: 89.

DuPraw, E. J. (1970). "DNA and Chromosomes." Holt, New York.

DuPraw, E. J., and Bahr, G. F. (1969). *Acta Cytol.* **13**: 188.

Goldmeier, E. (1972). Similarity in visually perceived forms. *In* "Psychological Issues," Vol. 8, No. 1. Monogr. 29 Intimate Univ. Press, New York.

Hamerton, J. L., Jacobs, P. A., and Klinger, H. P., eds. (1973). *Stand. Hum. Genet., Paris Conf., 1971.*

Hsu, T. C., Arrighi, F. E., and Saunders, F. G. (1972). *Proc. Nat. Acad. Sci. U.S* **69**: 1464.

McKusick, V. A. (1969). "Human Genetics," 2nd Ed. Prentice-Hall, Englewood Cliffs, New Jersey.

Schnedl, W. (1971). *Chromosoma* **34**: 448.

Schnedl, W. (1973). *In* "Chromosome Identification" (T. Caspersson and L. Zech, eds.), p. 342–345. Academic Press, New York.

Zech, L. (1973). *In* "Chromosome Identification" (T. Caspersson and L. Zech, eds.), p. 28–31. Academic Press, New York.

SUBJECT INDEX

CONTENTS OF PREVIOUS VOLUMES

Volume 2

A 4
B 5
C 6
D 7
E 8
F 9
G 0
H 1
I 2
J 3